Making Customer
Satisfaction Happen

Making Customer Satisfaction Happen

A STRATEGY FOR DELIGHTING CUSTOMERS

Roderick M. McNealy

Director of Customer Driven Quality at *Johnson & Johnson*
Hospital Services Inc.
New Jersey, USA

CHAPMAN & HALL
London · Glasgow · Weinheim · New York · Tokyo · Melbourne · Madras

Published by Chapman & Hall, 2—6 Boundary Row, London SE1 8HN

Chapman & Hall, 2—6 Boundary Row, London SE1 8HN, UK

Blackie Academic & Professional, Wester Cleddens Road, Bishopbriggs, Glasgow G64 2NZ, UK

Chapman & Hall GmbH, Pappelallee 3, 69469 Weinheim, Germany

Chapman & Hall USA, 115 Fifth Avenue, New York, NY 10003, USA

Chapman & Hall Japan, ITP-Japan, Kyowa Building, 3F, 2-2-1 Hirakawacho, Chiyoda-ku, Tokyo 102, Japan

Chapman & Hall Australia, 102 Dodds Street, South Melbourne, Victoria 3205, Australia

Chapman & Hall India, R. Seshadri, 32 Second Main Road, CIT East, Madras 600 035, India

First edition 1994
Reprinted 1996

© 1994 Roderick M. McNealy

Typeset in 11/13pts by Mews Photosetting, Beckenham, Kent
Printed in Great Britain by Athenaeum Press Ltd, Gateshead,
Tyne & Wear

ISBN 0 412 58920 6

A catalogue record for this book is available from the British Library

To Patty, Mary and Becky
whose love and support made this book happen.

Contents

Contents

Foreword

As a sales manager for Multimixer milk shake machines, Ray Kroc kept hearing requests from restaurant operators for 'one of those mixers of yours like the ones the McDonald brothers have in San Bernadino, California.' So Ray did some checking and found out that they didn't have one or two, but eight! He decided to see things for himself. During that first visit he was amazed at the crowds of customers. Ray's first action was to do what his instinct told him to do. He walked up to a customer and asked, 'What's the attraction here?' The customer replied, 'You'll get the best hamburger you ever ate for fifteen cents. And you don't have to wait and mess around tipping waitresses.'

From the opening day at his own McDonald's, Ray knew the customer was going to make him or break him. He applied his 'customer research' in formulating his company's focus on customer satisfaction. He called that focus 'Q, S, C, and V'; quality, service, cleanliness, and value. To this day, standards and measures for each of these are used to keep everyone in the McDonald's system focused on satisfying the 28 million customers we serve daily. We at McDonald's, view ourselves as having more than 14,000 customer satisfaction departments – our restaurants – all over the world which provide us with direct and immediate feedback on our business every day. Using several new customer feedback tools, restaurant performance is measured from the customers' point of view with improvements made accordingly.

The world is far different today than in Ray Kroc's first years at McDonald's, yet his strategic vision for the organization is as relevant today as it ever was and it is relevant to any type of organization. Public opinion and customers' needs and expectations are continually changing and evolving. There is a tremendous opportunity for those individuals and organizations that listen to and can understand and analyze what the consumer public is telling them.

We are now competing in a world of unparalleled opportunity and unmatched innovation. The speed of this innovation is also unmatched in any other era. Organizations can develop overnight and create entirely new segments or replace marketplace fixtures. These new organizations may represent entirely different skill sets and technologies, but their success is invariably due to one central factor. They are providing customers with a better product or service. Or a product or service that previously did not exist, or that existing market participants decided was unnecessary, inconsequential, or unimportant. It is this internally focused attitude that enabled those organizations focused on the customer and their needs and desires to build significant franchises and to do so at a remarkable pace.

Importantly, this 'formula' for success is not unique to any specific industry or occupation. The evidence is all around us, in every facet of our daily lives. We are always looking for improved products and services that do an increasingly better job of satisfying our ever-changing appetites. The key is to apply this personal life approach of always looking for something better to our professional life, no matter what that profession is. Our organization's customer is no different from us – they want better products and services too. So our decision is actually quite simple. We either decide to learn everything we can about our customer so that we can give them the product and service they want and more. Or, we can watch our competitors do it and then contemplate some new line of work.

Robin E. Johnson
Assistant Vice President
Quality Development
McDonalds Corporation

Introduction

Customer Satisfaction is the critical strategic weapon for the 1990s and beyond for any type of organization. Customer Satisfaction is so important because product and service companies alike, healthcare organizations, and even educational institutions must preserve their customer resources with the same energy and enthusiasm that we as a nation seek to preserve our natural resources.

Our customers are our organization's natural resources. We will be doomed as a civilization in spite of all our material accomplishments if we squander our natural resources. We face an equally bleak future as an organization of any type if we let our customer resources depart without any conscious effort to retain them.

The World Class Customer Satisfaction organizations are uniquely focused on their customers. This external focus is constant throughout the organization. There is no question in any employee's mind – from the chairman and chief executive officer to the newest member on the payroll – that customers are our most important resource, and therefore are to be treasured. Because our customers are so valuable, they should be an integral part of every aspect of our planning process. Their needs and expectations – even though they may be ever changing – should be continually monitored, measured, and communicated throughout our organization.

The World Class Customer Satisfaction organization realizes that everyone within it has a role to play in satisfying that customer. It is never seen as merely 'sales' job' or consigned to the realm of 'customer service' professionals. Everybody plays a part. The organization is structured to facilitate that role, not impede it. The added steps, the needless paperwork, the redundant management layers, the bureaucratic procedures have all been eliminated. Those in the organization dealing with our customer – and there are far more people within our organization that deal with our customer than we think! – are confident in the knowledge that the entire organization is focused to assist them in meeting and finally exceeding our customer's needs and expectations.

World Class Customer Satisfaction organizations reinforce their strategic and structural focus on the customer through their reward and recognition process. The message transmitted through a wide variety of employee recognition, reward, and promotion activities reaffirms the primacy of our customer – we will succeed when our customers' expectations are exceeded.

World Class Customer Satisfaction organizations are not limited to the Fortune 500, to 'industry', or to production and hardware organizations. Every organization has customers and it will succeed or fail based upon its ability to 'go beyond' what is currently expected. Clearly, automobile manufacturers, appliance and electronics makers, and gas stations all have customers. But so do hospitals, healthcare organizations such as managed care or Health Maintenance Organizations (HMOs), advertising agencies, banks, restaurants, hotels, school boards, college and university administrations, the military, and even our local, state, and federal governments.

Some of these organizations are acutely aware of the primacy of their customers and they will 'live to fight another day' because of their 'external' customer focus. Other organizations, from these same categories, have little or no understanding of their customers and they will be relegated to the dustbin with other organizations that failed to realize 'the customer is king, long live the king!'

In *Making Customer Satisfaction Happen* we are going to examine just what is meant by Customer Satisfaction, so that we have a common language for our common goal of exceeding our customers' needs and expectations. We are also going to examine in detail just why World Class Customer Satisfaction is such a vital strategic weapon for any type of organization. Next, we will examine case studies of organizations leading the Customer Satisfaction 'power curve' and understand what they are doing that could be applicable to any type of organization. Finally, we will examine how any organization can determine just what level of Customer Satisfaction they are currently achieving and to do so quantitatively.

The beauty of *Making Customer Satisfaction Happen* is that it is totally measurable. It is not a 'warm, fuzzy' feeling. We will examine how any organization can obtain a direct and unimpeded ear to the customer's voice and we should be ravenous to hear what they have to say.

As we enter the world of the twenty-first century, many new and different forces will confront us as individuals and organizations. Nothing is as certain as the constancy of change, yet some aspects of our world remain stunningly fixed. We must treat each customer as a valuable, irreplaceable resource and treat them as we would wish to be treated. It is just that simple. Others will endeavor to complicate, but common sense must prevail.

We must always consider how we would wish to be treated if confronted by your own organization. We must always remind ourselves that we must have consistent standards in our personal and professional lives. We are always seeking value, always seeking those goods and services which exceed our needs and expectations, so why should our customers be any different? Once we have embraced this 'single set of books' approach where our personal and professional behaviors are one, then we are well on the road to *Making Customer Satisfaction Happen* and we will see our customers in the following light:

A CUSTOMER

1. A Customer is the most important person in any business.
2. A Customer is not dependent on us. We are dependent on them.
3. A Customer is not an interruption of our work. They are the purpose of it.
4. A Customer does us a favor when they call. We are not doing them a favor by serving them.
5. A Customer is a part of our business, not an outsider.
6. A Customer is not a cold statistic. They are flesh and blood human beings with feelings and emotions like our own.
7. A Customer is not someone to argue or match wits with.

8. A Customer is a person who brings us their wants. It is our job to fill those wants.
9. A Customer is deserving of the most courteous and attentive treatment we can give them.
10. A Customer is the life blood of this and every other business.
11. A Customer is the person that makes it possible to pay our salaries.

Getting started

The customer is the most important part of the production line.
Without someone to purchase our product, we might as well
shut down the whole plant.

W. Edwards Deming

Examine the everyday products and services we interact with on a
continual basis and we will quickly see a multitude of examples of
customer satisfaction, or its absence. It is important to continually
focus on these personal, if mundane, examples because they are completely consistent with the approach to Customer Satisfaction that
we require in our professional life. Quite simply, the practices that
other organizations follow when they interact with us, and fail to
consider our needs and expectations as customers, are probably the
same practices that our organization follows when dealing with our
customers, be they other organizations or individuals. Let's examine
some examples.

The bank branch office

On a bright Saturday morning at about 10 am there are exactly
three individuals in the branch of a major northeastern American
bank. Two of the individuals are bank employees, one of whom is
the branch manager, and they are both standing in the teller's area
totally absorbed in conversation with each other. While the bank
customer, who is waiting to cash a check for some weekend cash,
is waiting, the teller and the branch manager have the following
conversation.

'I'll tell you, I'm just beat today. I hate working Saturdays!'.
'Me too, I don't know why we have to work Saturdays, I just
can't rest.'
'I know what you mean. If I was (sic) running this bank, we
wouldn't have Saturday hours. We're too crowded and it's just
too damned busy! Now, may we help you with something, sir?'

*Do our employees understand the value of each individual
customer?*

The drive-up window

The drive-up bank window is often seen as a major step towards
increased customer satisfaction. However, the drive-up windows at
a major bank do not open until 9 am, the same time the entire bank
opens and half an hour after most local residents have to be at work
or school. Additionally, while the bank's major operations close at
4 pm, the drive-up windows remain open until 6 pm, under the guise
of customer satisfaction.

But many people, particularly those who represent a large portion
of a bank's 'target audience', are unable to get out of work and to
the bank in time for the 6 pm deadline. Therefore, a service which
was implemented to increase customer satisfaction in reality only
serves to frustrate it.

In one case, a school teacher developed her own 'drive-up' bank
window system. She appears at the bank on mortgage payment days
at 7.30 am and slips her mortgage payment envelope through the
bank's locked glass doors to the custodian who prepares the bank
lobby each day. This custodian is the only bank employee whose
hours match this customer's. He cheerfully accepts the mortgage
payment envelope from her and sees that it is deposited and her
account credited. While this makeshift system works adequately
for paying the mortgage, it might prove a difficult way to apply
for one.

*Do our organization structure and hours of operation reflect
the needs of our customers, or are our maintenance personnel the
only employees whose hours match our customers'?*

2

The telephone contact

Mortgage rates are at a 30-year low and therefore many home owners are looking to refinance existing mortgages to achieve these lower rates, while new home buyers are 'taking the plunge' prompted by this 'once in a lifetime' opportunity. Yet, telephone almost any bank with the most basic information request concerning mortgage rates and we will be met with a response we might have anticipated for Jesse James, John Dillenger, or Willie Sutton – all of whom did not ask for a mortgage before taking money from banks!

> 'Hello, I'm calling to find out what your 30-year, fixed rate mortgage terms are for a refinance.'
> 'Just one moment, I'll connect you with the mortgage department.'
> 'Mortgage department, may I help you?'
> 'Hello, I'm calling to find out what your 30-year, fixed rate mortgage terms are for a refinance.'
> 'Oh, well I don't have those figures here. Now, was that a new home purchase?'
> 'No, a refinance of our existing home and your newspaper ad said to call about your great rates, so I'm calling and I'd like to know what these rates are.'
> 'Well, our loan officers are: 1. in a meeting right now; 2. in the field appraising properties this morning; 3. not in until 11. May I have someone call you back then?'

Does our organization make it difficult to conduct business with us?

On-the-job training

After particularly tough days at the office, Beverly stops at a local fast food restaurant and orders a cherry milkshake from the take-out window. While not great for her calorie counting, the milkshake is an invaluable psychological and therapeutic boost – a way she rewards herself for surviving another day of office firefighting – 'big business backdraft'.

Today, her order is taken by a bright and cheery newcomer to the restaurant service contingent. The newcomer puzzles briefly

3

while taking the order for 'one cherry milkshake', but she returns in ten minutes or so with the beverage. After taking Beverly's money and handing the milkshake over to her, the new employee waits anxiously for the results of the first sips.

'How is it, I mean I've never made one before and I was wondering if it tasted all right?' Instinctively, Beverly wonders why she had to pay full price for what was obviously a 'test case' and then she starts to re-think her plan for some elective surgery she was considering next month and begins to wonder what the term 'practising medicine' really means.

What message are we sending our customers when our front-line employees are ill-prepared for the task at hand, or are learning their jobs at our customers' expense?

Bait and switch

Bob feels almost guilty as he slips out of the office at 6 pm. It has been a particularly hectic day and the pace promises to continue into the evening as he needs to get home to get his daughter to her 4-H dog club meeting by seven. His wife has a full complement of term papers to correct for her seventh grade class and Bob's driving their daughter to 4-H frees up enough time for his wife to correct an extra four papers.

Bob starts his car without giving it a thought. Five years old and with 90 000 miles on it, it has never given the family a moment's trouble. Manufactured by a company renowned for product quality, Bob and his wife have already agreed that when the time comes for a new car, they will look at this model first.

Bob shifts the car into first gear, checks left and right, then eases out of the parking space and toward the parking lot exit. Suddenly, everything stops. 'Just not warmed up enough, good thing I wasn't out on the road', Bob thinks instinctively as he tries to restart the car. Nothing. Bob mechanically runs through his 'stall starting drill' by turning off the lights, fan, and radio, but his efforts to restart the car yield nothing. 'Well, this is why we joined the AutoClub', thinks Bob, but he realizes he actually preferred using their travel planning services rather than their roadside assistance program. Bob calls the AutoClub and home. So much for those four term papers his wife planned to correct with her 'found' free time!

Within half an hour a flatbed truck appears in Bob's office parking lot. There are few things more depressing than the sight of a flatbed truck, particularly with your car in tow. There is an air of finality about them. This was not going to be a case of merely jump starting a dead battery, or so it appeared. After a brief discussion of symptoms. the flatbed truck driver offered his diagnosis, 'Sounds like a timing belt and that's a big job.' Bob shivered and began sweating simultaneously. He could change the oil and filter in his car – and did so religiously every 3000 miles – but beyond that he was at the mercy of that great automotive subculture that appears to view the rest of society as ignorant rubes ripe for the taking.

Bob had the car towed to an authorized dealership. He had not developed one of those trusting relationships with a local mechanic who can work miracles that one often hears about around the office. The dealership would certainly be expensive, but they would know his car and he was willing to pay for the right job to be done on this previously trouble-free car. On the way to the dealership, Bob chatted with the flatbed truckdriver's eight-year-old daughter. She had come along to keep her father company and she carried on a non-stop conversation with Bob. She reminded him of his own daughter, now on the way to 4-H with her mother.

The driver had appeared distant after his initial diagnosis, possibly put off by Bob's pin-stripe suited office attire. Chatting away with an eight year old in the front seat of the flatbed truck, Bob now appeared quite 'a regular guy' to the driver. 'It's a big job, but not too big, maybe $300–$400.' Bob felt better – it's always nice to have a range of what 'a big job' meant, and this was a manageable range – not great – but manageable.

At the dealership Bob filled out the 'nighttime drop-off' card, as the service area closed at 5 pm. As the mechanic lowered his car from the flatbed, Bob completed the form with his name, daytime telephone number, and the amount he would authorize the dealership to spend without calling him first. Bob checked the box next to the figures '$300–$400' and thought, 'Let's hold it to that, if I check something higher, they'll just spend it anyway!' Bob was once again feeling himself at the mercy of the automotive subculture, particularly as the flatbed driver and his loquacious daughter drove off, leaving him alone at a closed dealership and waiting for his wife to drive him home.

5

The next day Bob called the dealership when it opened and clarified with the service manager the problems he experienced the day before. 'No problem, Bob, we'll call you back later this morning and tell you what we find.'

At 4 pm Bob had still not heard from the dealership and he could barely concentrate on the meeting he was attending, even though the company president was leading the discussion. Fortunately, Bob was not a major presenter, because he could only think that the more time that passed, the larger the problem they were finding at the dealership. Just as he was completing this thought, his secretary appeared at the door and told him the service manager at the dealership had called about his car. Instinctively, Bob excused himself from the meeting and called the dealership.

'Well Bob, we've got a problem here . . .' A cold sweat covered Bob's back as he sank into his office chair, his worst fears realized, and what did this overly familiar service manager mean '*We've* got a problem'? 'Our mechanic thinks there's major engine damage beyond the timing belt. He couldn't get the engine to turn at all and without taking the whole thing apart at $50 an hour, we think you should just consider a new engine for this car.'

This last statement was delivered so cooly by the service manager and with such finality that all Bob could think about was what *his* hourly wage rate was if the mechanics were making $50 an hour.

'What would a new engine cost?'
'About $4000'.

'What about a rebuilt engine?' Bob had heard that term and threw it out almost as a counter-offer to the $4000 proposition.

'We can put in a junked car engine for $1800'.
'What kind of guarantee does that come with, the junked engine?'
'30 days'.

Bob sat stunned. He told the service manager he would call him back tomorrow and make his decision. He told the service manager to find out if he really could get a junked engine because

6

$4000 for a car with 90 000 miles on it did not sound like a great investment.

Bob wanted to talk to his wife. She was always cooler in a crisis. The absolute amount of the dollars involved tended to blur Bob's mind, as he saw his options shaping up to be a junked engine in his 'no problems ever before car' or a new car, an option he had hoped to keep at bay for at least two more years.

Bob's wife recommended getting a second opinion. She reasoned that Bob's office had over 700 employees and invariably someone has that favorite mechanic. At least Bob could get a second opinion from someone like that. In the meantime, Bob's wife had already purchased the new car issue of 'Consumers' Reports' in preparation for the 'new car' option.

To his amazement, Bob found at least four co-workers the next day who had all had timing belt breakage experiences. Each told a similar story. 'No big deal, $300–400 in repairs, no need for a new engine or a junked one.' Each also had a mechanic they could recommend for Bob's 'second opinion'.

Bob called the mechanic who had been most enthusiastically recommended by his co-worker and described his experience. 'No problem, we do timing belts all the time. No need for a new engine. Just bring her in and we'll check for you. We'll arrange the tow and if you need a rental car to get you through the next few days, we have arrangements with a local rental car agency and can get you a $29 a day rate.'

Bob was almost speechless. His mind whirred over these simple statements from the mechanic. The mechanic's 'can do' attitude made Bob suddenly optimistic. Maybe it would be $400–500. Funny, that seemed like a lot of money when the figures were broached by the flatbed truck driver the previous day. Now, in comparison to $4000, it sounded like a bargain. And what about this tow truck and rental car offer? This mechanic was going to handle the whole thing.

The dealership offered only bad news and expensive solutions, yet this 'mom and pop' mechanic had the 'whole program' covered. 'Pretty impressive', Bob thought. He agreed to the mechanic's proposal and called the dealership telling them that he would seek a second opinion before going to the extreme of a new, or junked, engine or even a new car. Bob was rather surprised that the dealership's service manager was very calm at his request. The service manager said the car would be ready for the mechanic's tow truck the next afternoon.

The following morning, Bob received an unexpected call from the dealership's service manager. 'Bob, there has been something remarkable here.' Bob gulped, thinking he was about to be told his car had been stolen from the dealership's lot over night. 'Inexplicably, when our mechanic went out this morning to get the car ready for the tow truck pickup, the car started and ran perfectly. He had replaced the timing belt and it ran perfectly. I just can't explain it!' Just 24 hours previously this same man had told Bob his car was basically a worthless piece of junk with an irredeemable engine problem, and now it was running 'like a top'.

Bob's emotions and imagination had run the gamut in the past days, from the possibility of a new car for $15 000, a new engine for $4000, or that junked engine option with its brief 30-day guarantee that never really appealed to him. It seemed unlikely that the car would one day not be salvageable and the next day 'run perfectly'. Bob started to feel manipulated and a little angry. For the first time, he realized he could well have been scammed. He authorized the dealer to have the car ready for the tow. Bob wanted this 'second opinion' in spite of the 'miraculous recovery'.

One day later and Bob's latest impressions were confirmed by the 'second opinion' mechanic. 'Timing belt broke. Fixed it. Absolutely no other engine damage. We'd never replace the whole engine anyway. You do it in sections if you really need to, but your car doesn't. These cars are just getting started at 90 000 miles. You got another three to four years on this car. We gave it a complete tune-up and check-up. You'll need some rear brake work soon, but nothing serious. Do you know what the dealership was trying to do to you?'

This last comment caught Bob off-guard. He was standing face to face with a man old enough to be his father, a man who had opened his own business as part of a simple neighborhood gas station, yet stood before Bob in a clean blue uniform shirt with his name and his business' name embroidered above the breast pocket. Bob realized that all the mechanics working at this shop had on similar shirts and that they were all clean and freshly pressed. He remembered the dealership. Internationally recognized name, stylish interior, major highway location, service manager in a sports coat, tie, and clean hands. And they had tried to talk him 'out' of his perfectly good engine, which they could turn around and sell to some other trusting dupe while at the same time selling him a new engine, some junked engine of unknown origin, or best of all for them, maybe a whole new car!

> 'I think I do now. Thanks for being here. Give me some of your business cards, I'll put them up at the office. We've got 700 people there who would appreciate this type of honesty.'
> 'Sounds good to me. I can't afford to advertise, so I'd appreciate the "word of mouth". I think that always means the most anyway.'

Is our organization focused on short-term profits at the expense of long-term customer satisfaction and retention?

Each of these examples demonstrates the importance of Customer Satisfaction, or its absence, in everyday situations. We have probably all experienced something similar in the very recent past. But how does this apply to our organization? Maybe we are in charge of a school board, a large corporation, a non-profit organization, or an aircraft carrier. How do these examples relate to us? Our organization is completely different from every one of those situations just described. Or is it? In fact, every organization faces exactly the same issues and their inability to see these similarities is what keeps them from learning, growing, and succeeding to their fullest potential. So let's start with the very basics of *Making Customer Satisfaction Happen*. Let's start with Customers.

Throughout this book we are going to focus exclusively on external customers – the people who receive the goods and services our organization provides. We are going to single-mindedly focus on these external customers because they are the *sine qua non* of any organization – if we don't have customers, we don't have a business. Despite all the consultants' rhetoric, it is really quite a simple proposition. Our business or organization either satisfies it customers and does so on an ongoing basis, or we will not have a business or organization for long. So it is a question of focus and survival.

Many organizations have become involved in Total Quality Management efforts and have learned the importance of customers. They may even have adopted a definition of Quality that is determined by the customer – 'Quality is defined as meeting or exceeding the customers' needs and expectations', as we discussed in *Making Quality Happen*. However, some organizations have gone astray at this critical juncture in determining exactly who this customer is. Their rationale is as follows. 'Our organization has external and internal customers. Not everyone in the organization deals with external customers, but everyone has internal customers and suppliers. Therefore, the initial focus of our Quality effort will be on improving internal customer–supplier relationships. This way we can involve the entire organization at the start of our Quality Improvement effort.'

Although seemingly logical and praiseworthy, this approach is flawed in two important ways. First, the external customer is the reason our organization exists (Figure 1.1). Without them, we do

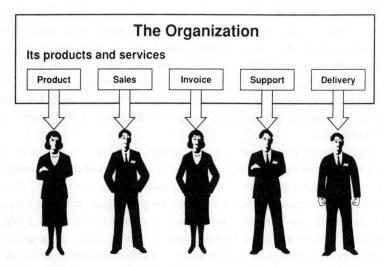

Figure 1.1 Definition: external customer.

10

not have an organization, and far more people within our organization come in contact with our customers than we currently understand, a point we will detail in Chapter 3. For now, just consider who in our organization 'touches' – comes in any kind of contact with – anyone from our customer's organization. It is probably a stunningly large number of people from a variety of disciplines within our organization.

The second flaw in the 'internal customer–supplier relationship first' approach is its internal focus. The entire reason why major organizations today are suffering, are losing market share, and are seeing their margins and absolute profits erode is because they are internally focused. By focusing only on our internal organization, we may well be spending valuable time and resources on activities of no value to the external customer. We may, in fact, be spending unnecessary time and effort on processes which merely add cost and not value to our actual products and services. Peter Drucker phrased it quite clearly when he stated, 'there is nothing so wasteful as doing well something that should not be done at all.' We can only determine what actually needs to be done when we focus on our external customer and seek to meet and exceed their needs and expectations. If we do not, we may find ourselves laboring under the *R.M.S. Titanic* School of Management.

The *R.M.S. (Royal Mail Ship) Titanic* was the industrial wonder of its time. Capable of carrying over 3500 passengers at levels of opulence even remarkable for the 'gilded age' in which it served, the *Titanic* was designed to cross the Atlantic from Southampton to New York in the then record time of under five days. On board the *Titanic* were two gyms, two swimming pools – a real novelty in 1912 – and a kitchen that produced 5000 rolls daily and carried 30 000 fresh eggs and 15 000 bottles of ale and stout in its stores. The *Titanic*'s officers were hand-picked from the best ships of the White Star Line, the premier Atlantic crossing line. The *Titanic* boasted a remarkable ship's orchestra that included some of the most accomplished musicians from the European continent [1].

Yes, all this and no binoculars in the 'crow's nest' – the ship's eyes in 1912, long before the advent of radar. The *Titanic* was a ship crossing the treacherous and icy North Atlantic at close to 25 miles per hour in all types of weather and trying for a record crossing

11

time besides, yet her management had decided to save money and not provide the crow's nest lookouts with binoculars. Even in 1912, binoculars were a standard crow's nest component in the North Atlantic trade routes, where it was not uncommon for steamships to plow under smaller fishing vessels and sailing ships on cold and foggy nights.

So, despite seven separate iceberg warnings from neighboring vessels, despite a captain with over 38 years experience in these same dangerous shipping lanes, the *Titanic* plowed into a black icefield at 22 knots and carried 1500 people to their deaths because some accountant at the White Star Line had decided to save the company $15 and not equip the crow's nest with binoculars. This accountant probably has a monument in the Accounting Hall of Fame, but the White Star Line is no longer in business [2].

The point of the *Titanic* School of Management is that if you lose sight, literally, of what is really important to the customer, such as actually arriving *alive* in New York from Southampton, all the other things your organization does are fruitless. The *Titanic* may have had a great orchestra, a marvellous kitchen staff, a capable and dependable purser's office, but they all drowned when the ship sank because the lookouts lacked binoculars.

This exact management approach is repeating itself in organizations around the world as you read this. Organizations are focusing internally, on *their* needs and expectations and *not* on their customers. Computer systems are developed and implemented to help the internal organization, not for their ease of use by the people who must actually deal with them on a daily basis. Insurance companies apparently pride themselves on complicated and obscure forms that baffle those whom the very insurance companies are supposedly serving. And, of course, the classic example is the American Internal Revenue Service that has publicly admitted it cannot develop forms that a citizen with a twelfth grade education can complete. The IRS is incapable of being clear, concise, customer-focused, and yet the very same citizens who must hire expensive accountants or tax preparers to wade through the nation's Byzantine tax laws are actually paying the salaries of the people who develop these indecipherable forms!

Therefore, throughout this book we are going to focus exclusively on the *external* customer, the recipient of our goods and services.

Importantly, this may not be one customer and this fact often 'throws' many organizations. In fact, we may have many customers for one organization. We may sell a medical instrument to a hospital based upon the recommendation of a physician or a nurse. Certainly, they are a customer. The actual device may be purchased and processed by a materials management staff within the hospital, so this group is also a customer. And finally, the money changes hands when our Accounts Receivables group interacts with the hospital's Accounts Payables staff, clearly another customer.

The critical point to remember is that each of these customer contacts is a customer and their needs and expectations need to be determined, and then we must organize our team to meet and then exceed these needs and expectations. Only when we have clearly focused on the external customer can we begin to work their needs and expectations back through our organization – the internal customer.

Importantly, every aspect of *Making Customer Satisfaction Happen* that deals with external customers is just as relevant to our internal customers as well. We just need to keep Edwards Deming's admonition clearly in mind as we focus our attention on our external customers. 'The customer is the most important part of the production line. Without someone to purchase our product, we might as well shut down the whole plant.'

WORLD CLASS – A DEFINITION

With our definition of the 'customer' clarified, we need to define what we mean by 'World Class Customer Satisfaction'. World Class organizations see Customer Satisfaction as an important, if not critical, strategic weapon in achieving their objectives. Clearly, these objectives vary by organization – to sell a million and a half automobiles this year, to have the leading marketshare in the disposable diaper category, to be the most highly rated school district in the northeastern United States – but what does not vary is their strategic focus on achieving these objectives through customer satisfaction (Figure 1.2). Invariably, these organizations have adopted a strategy similar to 'we will profitably achieve our business

Strategy:

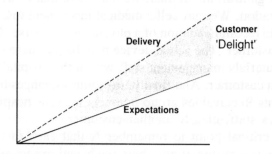

Retain current customers and gain new ones by meeting and exceeding their needs and expectations.

Figure 1.2 Definition: World Class Customer Satisfaction.

objectives by retaining current customers and gaining new ones by continually meeting and exceeding their needs and expectations in every part of our organization.' Almost any organization could adopt such a strategy because what varies by industry is the actual objective – the target result – not the strategy or process by which we will achieve our result. We will discuss this strategy throughout this book, but let's take a moment now to address frequently asked questions from skeptical corners by examining some key words that are included in this strategy.

The first of these words is 'profitably'. We are not proposing to 'give away the store'. Invariably, the skeptical will state that you can make customers satisfied by merely giving everything away, or by providing levels of service that are ruinously expensive. Therefore, we have included the word 'profitably' because we want to make it clear that we are in business to make a profit. The same thought process holds for non-profit or government organizations. While they may not be targeting to achieve a profit, they still must adhere to budgetary guidelines and live within those. So, a key part of our strategy is fiscal responsibility, no matter what type of organization we are.

14

The next key word is 'retaining'. Many organizations are continually hunting for new customers, new members, new students,
new recruits. The process of acquiring these new 'customers' is
invariably costly. We may have to advertise, 'Be all that you can
be in the Army', promote our product or service, 'for a limited time
only, buy three and get one free', or offer some other type of incentive to potential new members 'get one month free with our new
membership plan'. What compounds the problem is that we are not
acting in a vacuum, as our competitors are vying for these same
customers, which explains why our mailbox is full of advertising
circulars, our Sunday newspaper may contain numerous advertising sections and at least two sizeable coupon brochures, and why
14 minutes out of each prime time American television hour con-
sists entirely of advertising.

That is why we want to place our primary emphasis on retaining
current customers. Why? Because they are already our customers.
We have already gone to all the expense just described to obtain
them, so why let them get away? We will never recoup our initial
investment unless we get and hold our customers, and hold them
close to our hearts by the way. Merely by being our customers, they
have voiced a loyalty which we are bound to repay, so these current
customers must be our first concern.

We also use the word 'gaining' new customers in our strategy,
but only after we emphasize 'retaining'. We must continually be looking for new customers, new members, new students, etc. We do
not live in a static world and we are guaranteed nothing but the
inevitability of change. Those organizations that do not adapt to the
ever changing world will be doomed, in fact are doomed, to failure.
Previous success is absolutely, positively no guarantee of future
success. Just ask executives at International Business Machines,
General Motors, or the 18 members of the 1983 Fortune 50 that
were no longer on that prestigious list in 1993. So, we must target
new customers, but we must do so 'profitably'.

We next talk about 'needs and expectations' in our strategy. What
do these mean and aren't they different for every customer or potential customer? Certainly, needs and expectations do vary by customer
and most definitely by industry. However, there are very basic
customer needs and expectations that are consistent across industries

and those are at least in part typified in the examples at the start of this chapter. At the most basic level, we should treat all our customers as we would wish to be treated. Again, this is our central point of having one set of values for our personal and professional lives.

Invariably, all customers – regardless of industry – want to be treated as individuals who are important to our organization. Throughout the book we are going to examine organizations that understand this critical concept and put it into practice as the central focus of their organizational culture. These organizations actually believe in and live by the Definition of a Customer that appears in the Introduction.

Moreover, most customers want to be treated as individuals and want their needs and expectations to be seen as unique. Reflect on our personal experiences and we inevitably repel from situations where we are treated 'as just another number', possibly by a major utility firm, a large department store, or a credit card company. We want to deal with people, not machines, and people who are sympathetic to our situation. The wise organization understands this and, in fact, can use this personalized, individualized approach to its benefit.

Consider the case of L.L. Bean, probably the world's foremost catalogue sales organization. Call any time, day or night, and you will speak with an informed, interested, and intelligent customer service representative who will take your order, tell you the inventory status of each item, estimate the delivery timing, and continually refer to you by name throughout the transaction. And most of these telephone representatives are temporary employees at Bean, hired to cover the peak Fall ordering season.

The Bean organization has clearly understood the value of putting highly trained and carefully screened employees in direct contact with their customers – their first line of contact in fact! And they have clearly identified – probably quantitatively, as we will learn in Chapter 6 – the exact needs and expectations of their customers and those who would or could catalogue shop with them. This is not just a 'nice to have' on the part of the Bean organization. They are in constant touch with their customers and understand explicitly their customers' needs and expectations.

Importantly, it does not matter that our organization does not sell 'rip stop' pants or duck boots. What matters is that we understand the basic, customer-focused strategic direction embodied by the Bean organization. Frankly, if any part of our organization comes in contact with customers, and particularly comes in phone contact with our customers, then we have a great deal to learn from Bean regardless of how different our products or services may be.

Another almost universal customer need and expectation is 'value'. Many organizations mistake 'value' for 'price' and they try to compete merely on price cuts and deep discounts. In the absence of a direct, competitive alternative, this strategy can be successful. An example of this approach is the initial and relatively long-running success of K-Mart stores. Yet, when the customer is presented with an alternative that embraces more needs and expectations than merely price, and therefore starts to represent a better 'value', then the price factor becomes less important and less of a deciding factor in the product or service choice.

Value can best be seen as perceived Quality divided by Price:

$$Value = \frac{Quality}{Price}$$

If a product or service is the cheapest available, yet does not last or does not deliver as promised, then the majority of potential customers will move away from that product to those which offer greater quality at an equal or slightly higher price. Classic examples of this are the failure of the Yugo, Hyundai, and Renault Le Car models to be successful in the United States based on price alone. However, the GEO Prizm, Toyota Tercel, Ford Escort, and Honda Civic achieved excellent marketplace results, as slightly more expensive cars perceived by customers to possess infinitely greater value.

When we state Value in the form of an equation, one always runs the risk that somewhere there lurks a Quality engineer, or an accountant, who will try to draw Quality and Cost as potentially intersecting lines on a graph and look for that point of intersection at which we can maximize value with the minimum of cost. The secret is not in the graphing, but in the 'gabbing' – we need to be

17

continually talking to our customers and learning about their ever changing needs and expectations. That is really the role of 21st century marketing.

THE ROLE OF 21st CENTURY MARKETING

Most organizations are directed by a business approach we call '1960s Marketing'. '1960s Marketing' involves organizations hiring bright, creative – possibly slightly kookie – individuals to sit in conference rooms, or go on management retreats, and dream up new goods and services for our customers. These are organizations who use phrases like, 'let's run it up the flag pole and see who salutes', or 'if we throw 100lbs of spaghetti against the wall, some of it has to stick!' (I just would not want to be the next group to use that conference room!)

1960s Marketing is based upon the assumption that we can hire people who just 'know' what customers really want. How they actually acquire this knowledge is never specified. They apparently have some extra customer sympathetic gene or chromosome, but they nevertheless all seem to have 'it'. Most major advertising agencies for the longest time were dominated by creative groups who were perceived to have a great grasp of customers' needs and expectations. However, these creatives rarely ventured beyond the Hudson River and viewed anything beyond the Hamptons and mid-town Manhattan as 'Indian Country'. So much for customer contact. Their approach may have worked well in the 1960s and into the 1970s when the advertised products or services had little real competition. However, the tide turned dramatically in the 1980s and it continues to run out on this group into 1990s.

1960s Marketing really died in the mid 1970s when customer-focused organizations began to realize that the way to learn about customer's needs and expecations was to *ask* customers about them. And so customer-focused organizations went out and mingled with their customers. This mingling started with their management groups.

The management of World Class Customer Satisfaction organizations realized in the 1970s that they no longer had to hire creative people who instinctively understood their customers. Management

18

could cut out the middleman and go directly to the customer, speak to them first hand, learn their needs and expectations, and then *'mirabile dictu'*, actually develop the goods and services these customers were talking about.

The results were astonishing. Enormous business organizations appeared almost overnight because of their ability to listen and implement identified customer needs and expectations. Organizations such as Apple Computer, Sun Microsystems, Microsoft, Federal Express, WalMart are just some examples of organizations that grew at incredible rates to remarkable size based upon listening directly to customers and addressing needs and expectations established competitors had thought to be inconsequential.

Consider the fact that it took Sun Microsystems five years, Apple Computer seven years, and Microsoft 14 years to reach $1 billion in sales, while it took Toyota Motor Company 30 years, IBM 46 years, Johnson & Johnson 83 years, and Procter & Gamble 119 years to reach this landmark level. Clearly, there is nothing inherently 'wrong' with these latter organizations. However, there clearly exists tremendous opportunity in today's sophisticated business environment for even startup organizations if these organizations are uniquely focused on their customers and are actively and aggressively targeted at exceeding their needs and expectations to the point of delight.

LINKING TOTAL QUALITY AND CUSTOMER SATISFACTION

Now that we know what we mean by a 'Customer' and 'World Class Customer Satisfaction', how do they fit into our Total Quality Management process?

Most organizations that have been successful in implementing any type of Total Quality Management process have determined that there are three basic outputs of such a process (Figure 1.3). The single most important one of these is Customer Satisfaction. Customer Satisfaction is the key component in any Total Quality Management process. Customer Satisfaction, and in our discussion World Class

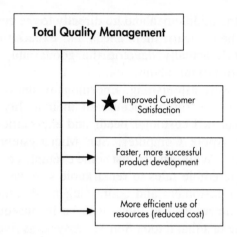

Figure 1.3 The outputs of quality.

Customer Satisfaction, means that we are delighting our customers by exceeding their needs and expectations.

By understanding our customers and their expectations better than our competitors do, we are better able to develop and introduce profitable new products and services specifically designed to exceed those expectations, the second most important aspect of a successful Total Quality Management process. We have just discussed some organizations that demonstrate the tremendous opportunity that exists for organizations that are uniquely focused on their customers' needs and expectations.

These organizations do not have to guess at what might succeed in the marketplace. They may not always be infallible, but their chances of success are far greater than those organizations still laboring under the delusions of '1960s Marketing'. This competitive edge is financially critical when it can cost a minimum of $35 000 000 to launch a consumer product such as a new shampoo or hand lotion. At that level of expenditure, we should seek any advantage available to improve our odds of success and our Number One advantage is our Customer and our knowledge of their needs and expectations.

By focusing our organization on providing the products and services our customers really want, we can reduce costs by eliminating those processes within our organization that do not 'add value' and that are not focused on the customer. This increased efficiency is the third major output of our Quality process.

These 'non value-added' activities are invariably the internally focused operations that take so much of our workday, yet have no relation to, or relevance for, our customer. Tom Peters has challenged the senior managements at many organizations to examine the content of their desktop 'in-boxes'. Invariably these are filled with memoranda and 'desk-drops' that focus on internal operations. Rarely, if ever, does any document relating to our Customer appear in this receptacle [3].

Yet the 'in-box' is the focus of our input from the surrounding environment, so if it does not appear there, it does not exist. Therefore, some organizations have found that their marketing departments, which should be the 'owner' of the Customer and their needs and expectations, spend as much as 30 to 40% of their office time on revising budgets and forecasts. Is this the best use of their time? Is this the type of activity that is going to launch the next breakthrough idea in our industry? This type of non-customer focused activity is the trademark of organizations that are internally focused and invariably playing 'catch-up' in the marketplace as a result.

'Reengineering' is a term currently in vogue to describe what is really the basic process of stripping away all the nonessential elements that have built up in any type of organization. Yet 'reengineering' is both puzzling and challenging many organizations simply because they do not clearly know what is nonessential – they do not know how to differentiate what is 'nice to have' from what is 'necessary to have'. And they do not know this because they do not know their customers. It all comes back to this very simple, basic tenet.

Once organizations strip away the unnecessary elements that encumber them, they will find that they are a more formidable force in the marketplace than ever, no matter what marketplace that might be. Importantly, these 'unnecessary elements' are rarely people. Unfortunately, much short-sighted 'reengineering' focuses on the 'quick fix' of 'downsizing' the organization to achieve a short-term financial impact. However, what remain after the people go are all the truly unnecessary tasks, procedures, and policies that have grown over time and are, in fact, strangling the organization. In reality, it is an organization's people who are the best source of information as to what is really necessary and what is needless. When we discard these people, we discard a great source of information on improving our organization from the inside as we prepare to build it by focusing outside.

SUMMARY LESSONS LEARNED

1. Adopt personal life standards for our professional life – have 'one set of books', not two. This avoids mental anguish and confusion.
2. View our organization as our customers view it, not as we on the inside view it. Would we want to confront our organization and deal with it on a regular basis?
3. Focus our efforts on the external customer, without them, there will be no need to consider improving our internal customer relationships.
4. Customer satisfaction is about exceeding our customer's needs and expectations to the point of 'delighting' them.
5. If we focus our organization on delighting our customers, we will be more efficient and effective, as we strip away organizational elements that do not add value to our customer relationships.

REFERENCES

[1] Lord, Walter (1976) *A Night To Remember*, Holt, Rinehart and Winston, New York, p. 120.
[2] Ballard, Dr. Robert D. (1987) *The Discovery of the Titanic*, Warner-Madison Press Book, Toronto, pp. 14, 18–29.
[3] Peters, Thomas J. and Austin, Nancy (1985) *A Passion for Excellence*. Random House, New York, p. 273.

Customer satisfaction pays   2

> Sell good merchandise at a reasonable profit, treat your customers like human beings, and they'll always come back for more.
>
> *Leon Leonwood Bean*

Now that we have clarified who we mean by the word 'customer' and what 'World Class Customer Satisfaction' entails, we should take a moment to outline the five basic tenets of World Class Customer Satisfaction that will be presented throughout the remainder of the book.

1. *World Class Customer Satisfaction is a critical strategic weapon that results in increased market share and increased profits.* Customer Satisfaction is not a 'program', a 'flavor of the month' panacea, or an adjunct to our Quality Improvement efforts. It is a strategic approach that our organization will use to achieve its critical objectives. While the strategic statement we presented in Chapter 2 is appropriate to almost any type of organization, the actual objectives of these organizations will vary widely.

    For example, Toyota Motor Company may have a business objective of selling 1.5 million automobiles this year, while Sun Microsystems may be targeting a five point market share gain in the workstation computer market. A major hospital chain may have as its objective to increase revenues and be recognized as the center of excellence for a specific medical procedure, such as hip replacements. A non-profit organization may have as its annual objective a 5% membership increase and full utilization of its numerous service components by its membership.

In each case, Customer Satisfaction is the strategic weapon which will enable these organizations to achieve their specific objectives. We will shortly demonstrate why we emphasize Customer Satisfaction as the primary strategic weapon. Clearly, it is not the only weapon, but it must be first among equals, as numerous organizations have demonstrated. They have also demonstrated that Customer Satisfaction 'pays', as reflected in the remarkable growth of organizations delighting their customers. The impact of Customer Satisfaction is not something that must be taken on faith. Its 'bottom-line' impact has been indisputably demonstrated and these results will be presented in this chapter.

2. *World Class Customer Satisfaction begins with its ownership by Top Management.* Because Customer Satisfaction is our primary strategic weapon, it must be identified as such by the leadership of our organization. All the organizations that are recognized as World Class Customer Satisfaction organizations have an unswerving commitment and ownership of Customer Satisfaction from their senior management. This ownership is visible, pervasive, and constant. There would be no question in our minds, as an employee in any one of the organizations we will be discussing, concerning this focus.

Managements in these organizations have realized that their actions – not their words – are what direct the behavior of their organizations and that is why the term 'ownership' is used in preference to words such as 'commitment' or 'support'. When we 'own' something, we are responsible for it, live with it, or suffer its loss. In cases where organizations have failed to satisfy their customers, that loss can be substantial, such as the $16.6 billion loss by IBM in 1993, the $2.6 billion loss by General Motors in 1992, Ford Motors' $2.2 billion loss in 1992 following its 1991 loss of $4.9 billion, or Chrysler's $538 million loss in 1991 [1]. This tenet will be discussed later in this chapter.

3. *World Class Customer Satisfaction involves the entire organization.* Management must 'own' Customer Satisfaction and make it their primary strategic weapon. However, they have the entire organization to assist them in this process and they must involve that entire organization if they want to maximize their success.

24

Far greater numbers of our organization come in contact with our customers than we currently realize. While we will detail this fact – and how to profit from it – in Chapter 3, it is important at this juncture to focus on the need to communicate management's ownership of Customer Satisfaction to the entire organization *and* management's enlistment of the entire organization in doing what is appropriate and necessary to achieve World Class Customer Satisfaction. This tenet will be discussed in Chapter 3.

Frederick Smith, chief executive officer of 1990 Malcolm Baldrige National Quality Award Winner Federal Express, continually reinforces both the primacy of customer satisfaction to his employees and their authorization to 'make it happen': 'The employee has to feel that he or she has the right, the authorization, or the backing to do whatever is necessary to satisfy the customer.' That type of management ownership and all-inclusive organizational involvement guarantees that the organization 'absolutely, positively' gets the message that 'the customer is king'.

4. *Customer Satisfaction has fundamental organization structure implications*. It is not enough that our organization understands our management's ownership of Customer Satisfaction, nor is it enough that it shares this enthusiasm. To be World Class, our entire organization will have to be structured to ultimately satisfy our customers. Invariably, all organizational structure theory is internally focused. Little thought is ever given concerning how our organization's structure impacts the ultimate customer. But just as many organizations underestimate how many of their employees come in daily contact with customers, so do they also underestimate how 'unfriendly' their organization's structure is to these same customers – and to the employees who represent our first line of contact with these customers.

Most organizations are structured as if their senior management were the ultimate customer – everything within the organization is focused on satisfying every whim of top management.

Some organizations have developed management information systems that are so sophisticated that they require only that the executive touch the computer screen to call up appropriate data or reports. New improvements on these systems will be voice activated, so the executive merely states their request to the computer and the information appears.

Contrast this internally focused innovation with the 'touch tone' nightmare customers often experience when calling our organization: 'If you are calling from a 'touch tone' phone, press "1"' etc., invariably followed by five minutes of touch-tone menus for the caller to wade through, at their expense. Who is this system really designed to help? Clearly not the customer who is paying our bills and paying for this dreadful and depersonalizing touch-tone menu system, whose installation was probably heralded internally as a tremendous cost saver.

However, the external customer determines whether our organization lives or dies and therefore the World Class organizations have realized that structuring their organizations to facilitate their customer interaction will ultimately help speed the customer cash flow that keeps the bills paid. They do this by keeping in touch with the customer, not through touch-tone phone menus. This tenet will be discussed and illustrated under a discussion of Customer Satisfaction Best Practices in Chapter 5.

5. *Customer Satisfaction can be quantified, measured, and tracked.* A great many management theories are just that – theoretical. They may sound good and even appear practical, but are many times hard to measure. Customer Satisfaction is ultimately so powerful because it is so quantifiable. The entire focus of Chapter 6 will be to demonstrate how any organization can quantify its current levels of customer satisfaction and then track their progress at continuously improving on the road to *Making Customer Satisfaction Happen*. Customer Satisfaction deals with data, not theory, and there is a beauty to data:

If we can define it – we can measure it;
If we can measure it – we can analyze it;
If we can analyze it – we can control it;
If we can control it – we can improve it.

And if you can improve it once, you can continuously improve it! We will discuss how this can be easily done in Chapter 6.

26

Let's examine the first two of these five tenets of Customer Satisfaction.

BOTTOM-LINE IMPACT OF CUSTOMER SATISFACTION

Customer Satisfaction has a tremendous impact on customer retention and repeat business. In fact, customer service problems leading to customer dissatisfaction are the number one reason why companies lose customers. These customer service problems are just like those presented in Chapter 1 and research by Quality Now, from a variety of industries, quantitatively demonstrates the negative impact this poor service has on our organization's financial health.

The 82% directly traceable to customer dissatisfaction with either the product or service dwarf any of the other reasons why customers leave. Significantly, the 68% service dissatisfaction level is almost FIVE times that of actual product dissatisfaction and it is more than SEVEN times that of PRICE, which is loosely translated here as 'lured by competition' (Figure 2.1).

Since most organizations tend to be internally focused, they believe that product differentiation will secure them a place in the customer's

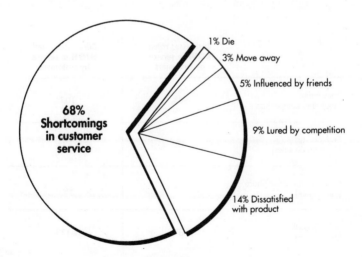

Figure 2.1 Why companies lose customers.

27

heart and wallet. When that differentiation is significant and clearly visible – as in the case of Apple Computers versus IBM personal computers – then they may be correct. However, in the vast majority of industries and organizations where the differences in products and services are less distinct – restaurants, banks, schools, retail stores, distributors – service plays the critical role in differentiating one from another, separating those that will win in the marketplace from those that will wither.

A clear, consistent, and constant communication with our customers will tell us immediately if they perceive any real difference in our product or price. If we have this knowledge – and we will discuss how to obtain it in Chapter 6 – then we will be able to pro-actively emphasize this product difference or we will, more likely, stress the service that accompanies the product. In either case, by being in close contact with our customer we have a distinct advantage in the marketplace that translates directly into a financial advantage.

Research by the PIMS, Strategic Planning Institute demonstrates the clear financial advantages of Customer Satisfaction (Figure 2.2). Organizations rated high in service by their customers: 1. enjoyed a 7% price premium to competition; 2. were 12 times more profitable – certainly helped in this case by the 7% price premium!; 3. were gaining marketshare while those rated low in customer satisfaction

	Businesses rated **LOW** in service by customers	Businesses rated **HIGH** in service by customers
Price Index Relative to competition	98%	107%
Profitability (% return on sales)	1%	12%
Changes in Market Share (per annum)	−2%	+6%
Sales Growth (per annum)	+8%	+17%

Figure 2.2 Benefits of high service levels.

28

were losing share; 4. achieved a sales growth rate twice that of their lowly-rated brothers and sisters [2].

Examples of organizations that are rated highly for service and are achieving these types of results are all around us. Apple Computers was able to enjoy a significant price premium to its IBM personal computer competition for many years because of its completely unique and customer preferred 'user friendly' operating system of icons and common sense. The Apple computer has been described as a computer designed by crazy computer 'chip heads' for humans, in contrast to the ponderous IBM hardware which was believed to be created by 'techies' for 'techies'. And while IBM was fighting clone computers – which consumers perceived as a comparable value to IBM from the outset of their personal computer introduction – Apple was able to avoid this price lowering competition through its major start-up period. It has only been the introduction of Apple emulating 'Windows' from Microsoft that has finally taken a bite from the magic Apple.

The Honda lawnmower, which we discussed in depth in *Making Quality Happen*, is another example of a relatively pedestrian product which, through superior customer focus, in the form of product and service, captured a major market share while charging a premium price in a market long dominated by household American names such as John Deere, Toro, Lawnboy, and Sears. The Honda lawnmower was developed by focusing on longstanding customer dissatisfaction with the ease of starting lawnmowers – they were just plain hard to start! By directly addressing and solving this issue – which competitors were aware of but disregarded – Honda was able to obtain the endorsement of customer-focused publications such as *Consumer Reports*, thereby economizing on its advertising budget, while still pricing its various models at the absolute top of the market.

Federal Express is another classic example that supports the PIMS findings. Federal Express started as the idea of Frederick W. Smith and its basic concepts were even part of an undergraduate paper he wrote at Yale. The paper received a 'C'. While we can laugh at that now, consider the environment of that time – the 1960s – and imagine our response to someone who would propose to charge several dollars to deliver 'mail' that we were currently paying only a few cents to send via the Post Office!

The key differentiating ingredient is the service aspect – Federal Express made it happen with next day delivery 'absolutely, positively' occurring almost 100% of the time. Given this service advantage, Federal Express was able to charge a premium and build a tremendously successful organization in the face of established market factors such as United Parcel and Emery. That success of course breeds competition, which Federal Express is currently facing.

However, the key learning from Federal Express is that an enormous organization was created by directly addressing customer needs and expectations, which existing organizations that should have been fully capable of addressing chose to either disregard, viewed as unimportant, or deemed not worthy of their consideration. When those situations occur and existing marketplace forces lose touch with their customers, the door is open for the bold, entrepreneurial forces that draw life from exceeding customers' needs and expectations.

Interestingly, the United States Postal Service awoke from its lifelong slumber and tried to offer a Federal Express clone in its 'Priority Mail' product. This product charges the customer $2.90 and targets delivery within two days. Unfortunately, recent studies both within the US government and by outside organizations have determined that the consumer actually has a better chance of achieving delivery within two days through regular US Mail at 29 cents than through 'Priority Mail'. Therefore, the Postal Service's failure to support this product with actual service will destroy any chance they had to compete with Federal Express.

Results from a Booz, Allen & Hamilton Inc. Strategic Planning Institute database study also reinforce the PIMS findings (Figure 2.3). This study demonstrates that organizations emphasizing Customer Satisfaction and customer service have 12 times the Return on Sales and six times the Return on Equity of their non-customer focused competitors. The study determined that by emphasizing Customer Satisfaction, these companies created increased brand loyalty and differentiated their products and services from the competition.

Investments in Customer Satisfaction were found to provide market share and financial benefits that compounded over time, whereas companies that emphasized cost efficiency over Customer

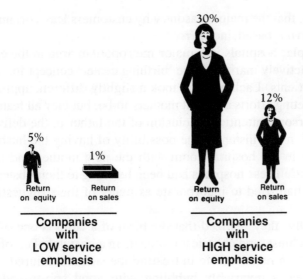

Figure 2.3 Benefits of high service emphasis.

Satisfaction achieved short-term profits and shareholder value, but not the longer term improved operating performance of those focused on Customer Satisfaction. Clearly, Customer Satisfaction pays, but who are those customers we are striving to delight?

TWO TYPES OF CUSTOMERS

There are two types of customers in a broad sense – ours and our competitors'. Often we disregard our competitors' customers for a variety of reasons, none of which are usually supported by data. For example, we disparage our competitors' customers by stating that they 'only buy on price' and are not sophisticated enough to appreciate the 'quality' our organization offers.

Importantly, our 'organization' in this example could be any organization and this rationale would hold true, whether we are a major corporation, a restaurant, a university, a non-profit organization, or a hospital. In fact, data have demonstrated, and we have already seen, that the major reasons why customers leave organizations are *service* based, not *price*.

For example, hospitals in a major metropolitan area in the early 1980s were actively marketing the 'birthing center' concept for their obstetrics patients. Each hospital took a slightly different approach to their marketing efforts aimed at mothers-to-be, but they all featured increased personal attention, inclusion of the father in the delivery process, and in one instance, the possibility of having the husband actually stay in the hospital room with the new mother and their 'blessed event'. These hospitals had been listening to their obstetrics patients and had tried to incorporate as many of their suggestions as possible.

Additionally, they realized that the birth of a child is one of the happiest, brightest, and cheeriest moments in a hospital. Too often, hospitals serve a reactive role in treating the sick and injured. The maternity area is invariably bubbling with good spirits and the enthusiasm of happy new beginnings. These hospital administrators realized this fact and the marketing potential of the maternity environment for future repeat business.

Specifically, a joyous and pleasant childbirth experience would leave a positive afterglow which can translate to future patient loyalty and repeat visits when the need for hospital care arises – 'They treated us so well at Good Samaritan Hospital when Tiffany was born, I'll always go back there'. This type of enthusiastic endorsement is hard to match and difficult to duplicate through advertising and merchandising.

However, not all the hospitals in this metropolitan area were changing their maternity care approach. One couple we interviewed explained that they had actually visited several hospitals in preparation for the birth of their child. Their obstetrician-gynecologist could deliver in any of the major community hospitals, so the choice of hospital was left to the prospective parents. (This is an early example of a doctor focusing on the patient's needs, rather than vice versa).

The couple was very impressed with each hospital and the special features each included in their maternity marketing efforts. Some of the items offered included a champagne dinner for the new parents, birthing suites in place of the standard sterile and apprehension-inducing delivery rooms, and the Le Boyer delivery method of low lights and a warm water bath for the newly delivered infant. Each hospital appeared to be trying to outdo the other in their efforts to attract maternity patients. Each hospital, except one.

The 'grande dame' of the metropolitan hospitals had a long and auspicious reputation for medical excellence. It had once been the most prestigious hospital in the community and the most sought after location for doctor's residencies. However, the passage of time had 'levelled the playing field' to some degree. New hospitals had been built with the latest equipment, expanded in-patient and out-patient hospital facilities, and had larger parking areas, thanks to their more suburban locations. Additionally, as newcomers in the marketplace, these more recently opened hospitals aggressively marketed their services to the community, something the 'grande dame' had never needed to do and which they looked upon with disdain and contempt.

Additionally, the 'grande dame's' obstetrics practice reflected their contempt for the 'patient focused' efforts of their more recently created rivals and provided an excellent example of hospital practices in the 1940s, not the late 1980s. When visiting the 'grande dame' hospital's maternity facilities, the young couple were informed that while 'natural' childbirth deliveries were performed, the hospital really preferred general anesthetic deliveries and certainly would not allow participation or even attendance by the father. This was a medical procedure, not a social occasion. Fathers were to wait in a father's waiting room, which resembled the smoke-filled scenes thought only present in the 1950s Hollywood comedies featuring Doris Day and an interchangeable cast of nervous, chain-smoking husbands.

Clearly, the 'grande dame' staff had no interest in, or inclination to cater to, their patients. The young couple came away from their visit with the distinct impression that the prospective mothers, and certainly their husbands, were viewed as bothersome nuisances who merely served to interrupt the highly structured and overly orchestrated procedures of this 'cathedral of healing'. Needless to say, the couple's son was born in one of the more 'patient friendly' hospitals.

Within a year of the child's birth, the couple read in the local paper that the 'grande dame's' maternity practice had all but disappeared. The newspaper report contained a detailed listing of the prestigious citizens born at the hospital, most from the first half of the century, interestingly enough. Hospital administrators from the 'grand dame' bemoaned the 'lowering of community health standards' that enabled lesser facilities to 'market' medicine and medical services. Clearly, even as the 'grande dame' was sinking beneath the competitive waves, her management – in a near classic example of the *Titanic* School of Management style – still failed to see the importance of focusing on patients, their particular and ever-changing needs, and excelling at the services that constitute such a large part of a patient's hospital stay.

This once outstanding hospital failed to recognize that its customer base and their needs and expectations were changing. The paradigm had shifted. The world was not locked in the 1950s. Those patients going to competitive hospitals were not less intelligent, less deserving, less informed. In fact, they were more demanding and better informed than their predecessors.

Therefore, we should never dismiss our competitors' customers. Rather, we should want to know why they are not *our* customers and what about *us*, not *them*, is keeping them from being our customers. In fact, we should know our competitors' customers as well as we know our own and we are going to discuss how to do that in Chapter 4.

OUR CUSTOMERS COME IN THREE TYPES

For the moment, we are going to concentrate on our customers. Our customers fall into three basic categories (Figure 2.4):

- *Dissatisfied* and in danger of defecting to the competition at any time;
- *Satisfied* and willing to stay until something better comes along;
- *Delighted*, with needs and expectations exceeded on a continual basis. These customers are loyal and singing our praises to the world.

The *dissatisfied* customer is just 'looking for the door' to leave our franchise. Unfortunately, they almost never leave quietly – much to the distress of our organization and our remaining customers – as we will shortly learn.

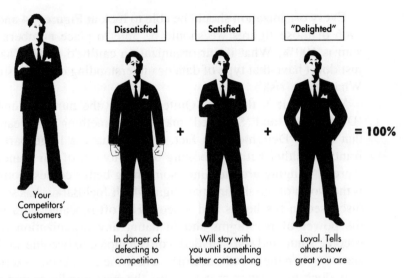

Figure 2.4 Three types of customer.

The *satisfied* customer is satisfied 'for now'. This is the trap that many organizations fall into on the subject of customer satisfaction. If they measure customer satisfaction at all, they take a very loose definition of the term 'satisfaction' and look at the 'top three boxes' when analyzing customer satisfaction ratings on a ten point scale. World Class Customer Satisfaction organizations, in contrast, only examine their 'top box' rating – they are only interested in knowing what percentage of their customers would give them 100% on their Customer Satisfaction exam. Many of their competitors lump their 80%, 90% and 100% ratings together in order to show their management higher scores, which only serve to deceive them, not their customers, who already 'know the score'! In Chapter 6 we will discuss developing accurate Customer Satisfaction vehicles in depth.

The *delighted* customer is actually a powerful marketing weapon for our organization and we will see in Chapter 5 how several organizations use their delighted customers as an integral part of their marketing effort. They have realized that delighted customers 'singing their praises' to other potential customers are an invaluable strategic weapon in their arsenal.

Every organization should be able to look at Figure 2.4 and under each of those figures we should be able to place numbers whose sum is 100%. What if our organization can't do that? What if we just don't have that type of data or understanding of our customers? What is the risk?

Well, what is the risk? Quite possibly the numbers under the 'Dissatisfied' and 'Satisfied' may total something far greater than that under 'Delighted'! In fact, we may have a large part of our franchise either actively looking for a way out of our franchise or merely 'hanging around' until something better comes along! This is the world of the shifting paradigm which Joel Barker has so clearly discussed in his books and videos. His oft-repeated phrase about the power of paradigms should haunt any organization that has become smug and self-satisfied – or is tempted to become so. Barker states, 'when the paradigm shifts, everyone goes back to zero!' [3].

It does not matter that you are the best candle maker on the continent if electric lights are now brightening our homes and offices. Nor does it matter that you are the best carriage maker on earth if everyone has fallen in love with the 'horseless carriage' of Henry Ford. When the world moved to chips and circuit boards, your great skills at producing vacuum tubes suddenly went for naught, and when our competitor – Motorola for example – is talking about 'Six Sigma' defective parts per million produced, our claims of X number per hundred fall flat.

Therefore, we need to know exactly what these numbers are for each of our three types of customers so that we can begin the process of moving the dissatisfied into the satisfied column and shortly moving both of them into the delighted category!

'WE DON'T GET MANY COMPLAINTS'

Many organizations are lulled into a false sense of confidence by the absence of any great number of customer complaints. In fact, many world famous organizations pride themselves on their 'customer satisfaction' programs that entail having an 800 (free call) telephone number on each of their packages to facilitate customer inquiries and then having an extensive 'customer service' staff on hand to handle these calls. The customer service department may be staffed with highly trained inviduals and many companies that deal with medical products or over the counter medications even have doctors and nurses on staff and available to handle calls.

These same organizations monitor and measure their calls. In several instances they even conduct statistical analyses on these measures to determine trends. Surely this is 'World Class' Customer Satisfaction. What more could an organization be expected to do?

This is an opportunity where we need to step back and re-evaluate our organization's actions in light of our own personal behaviors and standards. What would we do if we were confronted, as a customer, with our organization? The fact is that most dissatisfied customers do not complain. We can put 800 (free call) numbers on our products, questionnaires in our hotel rooms, even postage paid business reply cards on our home delivery pizza, and we will still not hear from the vast majority of dissatisfied customers, and the small number we do hear from can lull our organization into a false sense of security. Here is why.

The Technical Assistance Research Programs Institute (TARP) has conducted extensive research on this subject of customer complaint levels and their findings confirm what our personal practices should already have alerted us to (Figure 2.5).

The TARP research demonstrates that only 4% of dissatisfied customers complain. In most cases these 4% are located in mid-town Manhattan, New York, so if our organization is located anywhere beyond the Hudson River, we should not be depending upon customer complaint levels for our information on Customer Satisfaction.

Yet many large organizations do just that. Many product companies wait to hear from customers, by phone or mail, and even then disregard these inquiries or complaints unless a statistically significant number appears. Politicians wait to hear from their constituents after they have floated their latest 'trial balloon' for some new program, or more likely, some new tax. They wait to see their mail and check their telephone responses before determining whether to pursue their stated goals or backtrack.

The Clinton Administration in the United States has been notably active in the process of government by 'trial balloon'. Administration

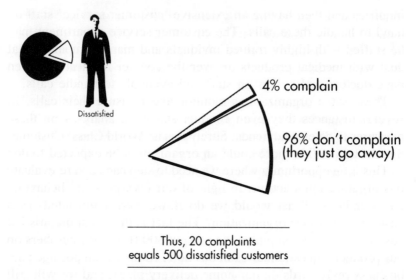

Dissatisfied

4% complain

96% don't complain
(they just go away)

Thus, 20 complaints
equals 500 dissatisfied customers

Figure 2.5 The nature of dissatisfied customers.

spokespeople continually 'leak' policies under consideration to the news media in order to evaluate public and congressional response. Notable examples of this approach were the 1993 National Budget, the Health Care Reform program, the funding for the proposed Health Care package, Bosnian intervention, and even the highly controversial subject of gays in the military.

Each of these was floated as a 'trial balloon' in a variety of formats by the Clinton Administration. The subsequent proposals created by the administration were invariably far different from the 'trial balloon' based upon the dramatic public response. Importantly, this response not only came in the form of telephone calls and mail, but also through constituent contacts with members of the United States Congress. In fact, the White House and Congress received a record five million phone calls from constituents during the five days following President Clinton's August 1993 budget speech, the majority of the calls being against the proposed budget [4].

The majority of these calls were against the President's proposed budget and the one vote majority passage by the United States Senate reflects their terror of constituent retribution. In fact, most congresspeople who voted for the budget were terrified of the prospect of the summer congressional recess when they would have to return and face the voters in their home districts. That direct constituent contact would demonstrate how dramatically understated even the statistically significant phone and mail response had been.

Here is an example where organizations have response rates covering millions of 'customers', and even these responses are understated. Consider the potential traps other organizations confront, for example in the health care industry. Many hospitals conduct surveys of patients as part of their check-out procedure. Still others subscribe to national or regional hospital surveys which combine the findings of a large group of hospitals.

What is the problem with this approach? Isn't any type of customer input better than none at all? In fact, the customer input that many organizations receive is misleading at best and inaccurate at worst. Even the best hospital conducted surveys are based upon a 30% return rate – out of all questionnaires distributed, 30% are returned. In fact, a 30% return rate is excellent for most market research, but this response rate also immediately demonstrates the flaws of this type of research.

Real scientific market research is conducted among closely monitored samples where even minor trends can be projected. It is possible to project major national trends through a closely monitored and scientifically designed sample of 1000 or less. In fact, the famous 'Willie Horton' advertising commercial used in George Bush's 1988 election campaign was based upon research with less than 100 participants.

However, the 30% of our patients who return their questionnaires is anything but a scientifically constructed sample, This is another opportunity for us to examine our own personal behaviors. If we were checking out of a hospital, would we take the time to fill out a questionnaire for an organization that has just charged us anywhere from $600 to $1000 a night? How many of us fill out the omnipresent questionnaires in hotels? That environment is invariably far less stressful than that found in a hospital. Therefore, we should not assume that any substantial number of our customers are going to fill out these response surveys or cards. We should not be expecting our customers to do our work for us, which is exactly what we are doing in a 'passive' customer questionnaire process.

39

Yet many hospitals and a variety of other organizations – from hotels, airlines, to national pizza chains – depend on these surveys and telephone contacts for their main customer input. We are not hearing from many dissatisfied customers, so our organization probably doesn't have a problem. Or does it?

DISSATISFIED CUSTOMERS DO NOT GO AWAY QUIETLY

Only 4% of all our dissatisfied customers complain, so 96% of our dissatisfied customers just go away. The problem is, they do not go away quietly. The landmark TARP data reports that the 96% who go away without telling us go out and tell from 10 to 15 other people about their bad experience with us. There is a group within this 96% – a 'lunatic fringe' of 13% – that will tell 23 or more people about their bad experience with our product or service (Figure 2.6).

Again, consider our own personal behavior. We have a bad experience with a product, a service, a bank, a restaurant, a movie. More than likely we will not tell the provider of that product or service, but we certainly will tell our family and our co-workers.

Dissatisfied

Each of the 96% that 'just go away' tells between 10 and 15 other people about their bad experience with your company's product or service.

Thus, the 480 non-complainers tell more than 5 000 others about their bad experience. 5% (250) will be influenced.

Figure 2.6 Complaints: the tip of the iceberg.

And bad news travels fast. We tend to be much more likely to believe bad news than we will favorable comments about a product or service. This is just a reflection of our increasingly skeptical society. We have been pummelled by superlatives – greatest, softest, fastest – to the point where we, as a society, are numb to these 'come-ons' and are therefore very receptive to any news that deflates the balloons created by publicists and advertising.

We are very likely to tell several co-workers about our bad experience and we have all been in group situations where one individual will really 'get rolling' and tell us one horror story after another about some terrible experience with some product or service they've encountered. Just ask the owner of a Jaguar, a Dodge Aspen, or a Plymouth Volaré – the latter two being the most heavily recalled cars in United States automotive history – about their cars, if you have several hours to spare! So that 'lunatic fringe' of 13% who tell 20 or more people is probably someone we have come in contact with and very likely in the not too distant past!

If these dissatisfied customers are telling 10 to 15 other people, that would be bad enough. But do we imagine that the telling remains the same from person number 1 to person number 15? In fact, as we all know from personal experience, the story grows and worsens as it is told. When we tell person number 1 about our bad meal at Maison de Fred we might mention the inattentive service and expensive bill. By the time person number 15 hears the story, we were taken out behind the restaurant and beaten by Fred and had our parentage questioned by the headwaiter!

THE 'STEALTH' COMPLAINT LEVEL

The actual mathematics of this situation are terrifying, particularly if we are not now even tracking the complaints from people who contact our organization. If our organization only receives 20 complaints over whatever time period, we more than likely actually have 500 dissatisfied customers. The problem is that we do not know anything about the 480 that have not bothered to contact us. However, TARP has now informed us that these 480 – along with the 20 we heard from – are telling more than 5000 other individuals about their

41

bad experience with our organization. The research also demonstrates an even more terrifying fact. Five percent of these 5000 contacts – 250 individuals – will be seriously influenced by what they have heard negatively about our product or service. Seriously influenced enough to consider not buying or even trying it.

Therefore, the 20 seemingly harmless complaints that we received have transmogrified into 500 actual complaints and encompassed an additional 250 individuals who 'hate' us, yet have never even met us! Our 20 complaints actually represent 750 dissatisfied customers – or potential customers – of whom 730 are completely unknown to our organization.

Another sobering fact is that 20 complaints to any type of organization over any reasonable time period – one week, one month, one quarter – would not be viewed by most managements as alarming. Check your organization's complaint levels today and you may be surprised to find that you are receiving this many complaints daily!

CUSTOMER SATISFACTION 'QUICK FIX'

In *Making Quality Happen* we introduced a five-step Problem Elimination Model whose second step was a 'Quick Fix'. The objective of this step was to 'patch' an existing problem to keep the organization afloat while concurrently searching out the 'Root Cause' of the problem so that it could be eliminated, and with it, the problem itself. The problem with most 'Quick Fixes' is that they have the nasty habit of becoming 'permanent' once we have stopped the 'bleeding'. We quickly forget that our original objective was to search out the problem's root cause. Instead, we are off to slay another organizational 'dragon' and we leave the 'Quick Fix' in place. In the case of our 20 complaints, there is some good news for our organization.

The TARP data states that of the 20 who complain to us by phone or letter, 60% will remain our customers *if* we resolve their problem to *their* satisfaction. Ninety-five percent of the 20 will remain our customers if we resolve their problem *quickly* and to their satisfaction (Figure 2.7).

Thus, the good news is we can keep between 12 and 19 of the 20 who originally complained. We have been able to do this because

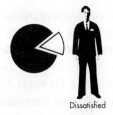

Dissatisfied

Of the 4% who complain, more than half
(60%) will stay if their problem is resolved.
95% will stay if their problem is resolved
quickly.

Thus, of the 20 complainers, you can keep
12 to 19 with a responsive complaint resolution
program ('hot lines' and warranties).

Figure 2.7 Complaint resolution.

our organization had an 800 telephone line and/or customer service
department, which are really just reactive customer satisfaction
vehicles. Many consultants have stated that it is, in fact, good to have
problems because we then have the customer's complete attention
and the opportunity to demonstrate to them our total commitment
to their satisfaction. While it is true that we will certainly have their
complete attention, this type of 'high wire act' management style
is extremely risky, particularly if our competitors are exceeding our
customers' expectations as a routine matter of course. In fact, our
reactive flailings may merely appear comical to our customers.

Consider the case of the advertising campaign for the General
Motors Saturn car. In Chapter 4 we will discuss in greater detail
the process for understanding customers' needs and expectations that
enabled Toyota to develop the Lexus cars in two years while General
Motors took seven to develop the Saturn. For now, let's examine
some of the initial advertising that appeared to promote the long
awaited Saturn.

Some of the introductory print and television advertising featured
a school teacher in Alaska who buys a Saturn sight unseen because
she believes in General Motors and their pledge that the Saturn car

is the beginning of a new commitment to the car-buying public. The thrust of the commercials and print ads is that a 'new day' was dawning in Detroit and that Saturn-buying customers could depend on General Motors.

A follow-up commercial featured the trials of a General Motors customer service employee as he endeavored to deliver a replacement front seat to this school teacher in the wilds of Alaska. The commercial includes humorous vignettes of the serviceman squeezed into commercial airliners with the replacement seat next to him and then finally arriving at the school teacher's home in the seaplane reminiscent of the beginning of the movie *Raiders of the Lost Ark*. No doubt the television commercial writers and producers believed they had crafted another touching masterpiece – *Gone With the Wind* in 60 seconds – demonstrating the warm, loving side of General Motors, the company with the 'caring Quick Fix'.

However, what message are they really conveying in this commercial? What they are actually stating in this commercial – at a cost of probably several hundred thousand dollars in commercial development and production costs, and several hundred thousand more to purchase the media in order to air the commercial – is that General Motors could not build the Saturn car correctly, even though it took them three times longer than Toyota took to produce the Lexus or Honda took to introduce the Accura! One hopes the executive offices in Detroit are on the ground floor because any executive who approved that advertising strategy should either be on the window ledge outside their office or be signed up for an upcoming course in Hara-Kiri with the new, improved Ginsu knives!

If we are satisfied with a Quick Fix approach to customer satisfaction, we are entirely missing the major issue, particularly in terms of raw numbers of customers. True, we can keep 19 out of the 20 customers who complain to us if we are appropriately subservient and speedy. However, in this Quick Fix mode we are entirely oblivious to the 730 *other* dissatisfied customers who will possibly never consider our products or services again.

Much to our chagrin, and our competitors' delight, we focused on the 20 dissatisfied customers we heard from, while the initial 480 dissatisfied customers who did not complain are gone because we never had the chance to learn of their problems or even try to solve them. So reactive, or Quick Fix, measures are just not enough. We need to take a proactive stance towards our customers.

THE DELIGHTED CUSTOMER AS AN ASSET

The same TARP data that tells us all the bad things that can happen to our organization through dissatisfied customers also tells us that the 'delighted' customer can have a significantly positive impact.

The 'delighted' customer tells 5 others and will have some degree of influence over them. Therefore, 'delighted' customers are a sound investment and some companies clearly understand their impact.

For example, Toyota has positive 'word-of-mouth' advertising from satisfied customers as one of its three strategic approaches for building their business. Discuss Nordstrom's Department Store, SAS Airlines, L. L. Bean catalogue shopping, even the famous Stew Leonard's dairy stores in Connecticut, and we will hear one great story after another from satisfied customers. What would we pay to have this type of positive, enthusiastic public relations? Do our customers feel this way about us? Would we even know if they did?

'Delighted'

The 'delighted' customer has received an unexpected service from the company. He'll tell 5 other people how terrific you are. 5% will be influenced.

Thus, delight 100 customers who will tell 500 others. Two dozen may become new customers.

Figure 2.8 The 'delighted' customer as an asset.

45

If we do not know how our customers feel about our organization, an examination of the customer satisfaction mathematics is sobering (Figure 2.8). While our 20 dissatisfied and vocal customers actually represent a potential 750 seriously displeased customers, 20 'delighted' customers are only going to translate to 25 total 'delighted' customers (the 20 delighted each tell five others and 5% of this total audience of 100, or five individuals, will be influenced). So the organizational impact of dissatisfied to delighted customers is 30 to 1! This relationship alone should jolt our organization from a reactive to a proactive Customer Satisfaction posture.

THE LIFETIME VALUE OF ONE CUSTOMER

The TARP data should cause any organization to look at its customers in a new light. They should not be seen as some nameless, faceless mass – the 'them' so often referred to in organizations. Customers have to be seen as individuals just like us. That is why we are continually emphasizing the need to personalize Customer Satisfaction – how would we react to our organization, how do we react when put into situations that we commonly place our customers in?

If the TARP data appear too impersonal, consider what one individual delighted customer represents to our organization. 'Delighted' customers are those who share their positive experience with others. This 'delighted' customer is a *retained* customer and certainly not part of the 90% of our dissatisfied customers who are *actively* looking for alternatives to us.

This retained customer is the key to marketshare growth. Earlier in this chapter we saw how organizations which satisfy their customers have marketshare growth rates far exceeding those of organizations which do not emphasize Customer Satisfaction. These Customer Satisfying organizations achieve these growth rates because they maintain and build their customer base. They do this because they understand the *value* of a retained and satisfied customer.

Once again, consider our own personal experience. Select one product or service that we utilize on a frequent basis. We might consider our bank, stock broker, auto repair shop, bakery, grocery store, pharmacy, pizza take out, or lunch time fish and chips shop.

What is our value to that organization? Do we think they know, or care? World Class Customer Satisfaction organizations certainly know and care and, in fact, can tell us directly and quantitatively what we represent to their organization over a lifetime. Yes, over a lifetime, because that is how World Class Organizations view their customers.

Ordinary organizations fall into the trap of seeing their customers as an indecipherable and indiscriminate 'mass' utilizing our product or service on a 'one off' basis. Conversely, organizations that emphasize Customer Satisfaction and customer retention understand the long-term value of a customer. This is not just so many words. The hard, cold, long-term value of a retained customer can be calculated in each of our organizations. Here is how some organizations have calculated the value of their customers.

Our neighborhood pizza shop should know that the frequent pizza eater averages one large pizza per week, at an average charge of $10. That translates to $520 per year or $5200 over a ten-year horizon. If we factor in the TARP data and assume that this 'delighted' pizza customer tells five others about our 'extra cheese and mushroom' special and that 5% of these contacts will at least try us, this adds an additional $130 per year for that customer or $1300 over our ten-year horizon. Thus one satisfied customer for our pizza shop is worth $650 per year or $6500 over ten years, and that's just from one pizza per week. What am I worth to you if I have a party and decide to serve pizza or if I decide to treat the sixth grade class as part of a school trip?

The automobile industry believes that a loyal customer represents a lifetime average revenue of $140 000. Why then, do some companies, like Lexus, go out of their way to 'delight' customers while others do not? If all automobile companies really saw each individual customer as $140 000 why would it be necessary for states within the United States to initiate 'Lemon Laws' which require various levels of mediation by the manufacturer if the customer has the same problem on a repeated basis.

In one classic instance, a husband and wife team of lawyers bought an internationally recognized make of car, Brand Y, after falling in love with its luxurious interior and sporty acceleration. Unfortunately, the car also came with a nasty habit of dying on the highway for no apparent reason as part of its standard equipment package!

After a long and litigious ordeal, the husband and wife rid themselves of the car and the manufacturer agreed to refund their purchase price. Several months later the couple received an out-of-state phone call. The caller confirmed the couples' name and inquired if they had ever owned a sporty Brand Y car. The lawyer couple stated that yes they *had* owned one. The caller then continued and asked them if it had a habit of dying on the highway with potentially terrifying consequences. They averred.

Both parties soon realized that the manufacturer had taken back the car from the two lawyers, transported it to another state without a 'Lemon Law', and then resold the car to an unsuspecting, non-lawyer customer without removing the original service manual from the car. It was this service manual, with the original owner's name still in it, which enabled the car's second distraught owner to contact its initial dissatisfied owners.

This is clearly the *modus operandi* of the short term, myopic organization that grabs today's profits at the expense of tomorrow's, next week's, next year's. Interestingly, Brand Y decided that they would no longer sell their cars in the United States within a few years of this incident, which was probably not 'one of a kind'.

Stew Leonard in the grocery industry sees each shopper as worth $50 000 over a ten-year period: $100 a week, 50 weeks a year for ten years. Therefore he gives their children free ice cream cones and will not allow lines at the check-out. People will drive out of their way to come to his stores when major grocery chains are far more convenient. What does he know that his competitors do not?

Do you think Procter & Gamble knows the lifetime value of a satisfied and retained sanitary protection product customer? This is a $2 billion category in the United States alone, whose target audience is every woman from the age of approximately 12 to 49. Market studies have shown that the average female customer will utilize a total of 6800 sanitary protection products in a variety of forms during the years from 12 to 49. This translates to each customer being worth $650 to a manufacturer of these products. Importantly, every woman from the age of 12 is a prospective customer, so the multiplier effect of maintaining women in a specific company's franchise is enormous.

Wouldn't it make sense if the manufacturers of these sanitary protection products for such a valuable customer listened to what that customer wanted? Yet, for many years customer needs and expectations were subordinated to the requirements of the engineers who designed the machinery in this capital intensive market. Marketers in this category were told by machinery loving engineers to sell what the machinery could make, rather than have the machinery tailored to produce what these valuable female customers wanted. One has to wonder how many, if any, of these inflexible engineers were female.

There was also tremendous financial pressure to maintain the 'status quo'. The Toxic Shock Syndrome scare had a devastating impact on the internal – tampon – sanitary protection category, so that producers of external products were swamped with incremental demand. With this extensive demand, it became very easy to rationalize not making the product improvements customers had been requesting in market research studies. Follow the logic of the managements in place at that time. They were already successful and now an unforeseen event has made their external product even more popular! Why 'upset the apple-cart'?

In fact, if they had really been listening to the customer – instead of their bankers and stock analysts – they would have learned that their customers were merely 'satisfied' with the current products. As we discussed earlier, this meant that customers in their franchise would remain with that product until something better came along. It is the very fact that self-satisfied managers allowed unmet customer needs and expectations to continue that enabled competitors to gain entrance to this enormous market.

Existing manufacturers in the sanitary protection category conducted studies among women that sought to determine their needs and expectations for these products. This research clearly enumerated these wants and needs. However, when the research was presented to management the decision was made to continue with the existing product. This product was selling exceptionally well and any changes, even if they were improvements that customers wanted, would be costly and time-consuming. This cost and time factor would have a direct impact on the organization's short-term financial performance and this limited horizon financial measure was exactly how this management was being evaluated by its senior management. Therefore, the stage was set for anyone bold enough to address customers' needs while holding the bankers and engineers at bay!

When a manufacturer finally entered the market with a product that began to address some of these pre-existing customer needs and expectations, customers quickly transferred their allegiance to that newcomer and therefore Procter & Gamble was able to secure a 30% market share of dollar sales in 1993, while just ten years before they had not been a factor in this market!

Or consider a similar product, disposable diapers. This is a $4 billion market in the United States alone, where the average infant uses a total of 7650 diapers from birth until the age of approximately three. This translates to each infant representing $2000 in diaper sales to manufacturers.

Is it any wonder that Procter & Gamble started paying closer attention to infants and their mothers, as opposed to inflexible engineers, when their 50% of this market faced challenges from both other United States manufacturers and the spectre of a revolutionary diaper under development in Japan? Evidence of this newfound customer focus is the rapid testing and national introduction of a wide variety of diaper products aimed at specific market segments, including boy-girl diapers, training pants diapers, and elastic leg and waist diapers. Having seen their own success in the feminine sanitary protection market as an example, Procter & Gamble was unwilling to watch their $2 billion stake in this market diminish without a fight over each one of these $2000 infants.

Armed with the quantitative knowledge of what each customer is worth, aren't our competitors better able to be even more efficient and effective marketers? Can we really allow our competitors this advantage?

Here again is an example of where it is necessary, and exceedingly profitable, to challenge existing paradigms. Some marketers competing with Procter & Gamble felt the elastic leg and waist band diapers were 'fads', that customers would not pay a premium price for such products. This was management's paradigm, not the customer's perception. If these executives had been listening closely to the customer, they would have heard the customer's cries for a better fitting diaper. Yet, how many of the executives who made

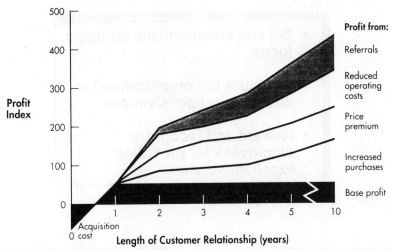

Figure 2.9 The value of one customer.

those decisions actually ever met with large numbers of mothers with infants? How many had infants in diapers? How many had ever held an infant with a leaking diaper?

Therefore, it is critical to view our customers over the long-term and quantify their value to our organization. Figure 2.9 presented by Frederick F. Reichheld and W. Earl Sasser, Jr. shows the profit potential of a 'delighted' and retained customer [5].

How profitable are we going to be if we are always looking for new customers to replace those we have dissatisfied? Look at the incredible pain, financial loss and expense American auto makers are experiencing in order to get the American auto buyer to even consider buying an American car.

Therefore, this concept of the 'lifetime' value of a customer directly addresses our second key tenet of *Making Customer Satisfaction Happen*, that management must 'own' customer satisfaction.

TOP MANAGEMENT MUST OWN CUSTOMER SATISFACTION

Making Customer Satisfaction Happen has significant organizational implications because of its strategic nature. Strategic issues are

> ★ **Set and communicate strategic focus**
>
> ★ **Structure the organization for customer responsiveness**
>
> ★ **Reward and recognize employees for providing excellent service**

Figure 2.10 The role of management.

management owned and therefore management plays a critical role in *Making Customer Satisfaction Happen*. We used the term 'ownership' of customer satisfaction in place of terms such as 'support' or 'commitment'. Management owns everything in our organization that is important – sales, profits, earnings per share, budgets, membership levels, etc. It can be no different with customer satisfaction.

Management has three basic ownership roles in institutionalizing Customer Satisfaction. The first is to make Customer Satisfaction a strategic focus of the organization. The second is to structure the organization to make it more responsive to the 'Voice of the Customer'. The third is to align the organization's reward and recognition system with its customer focus (Figure 2.10).

SET AND COMMUNICATE STRATEGIC FOCUS

Customer Satisfaction is a strategic approach to achieve our business objectives. The best example of this strategic approach is Toyota Motor Sales. (This example is detailed in the Best Practices Examples in Chapter 5.)

The 1992 business objective of Toyota Motor Sales was to sell 1.5 million automobiles (Figure 2.11). They strategically planned

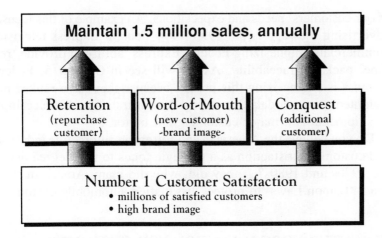

Figure 2.11 Strategic approach: Toyota.

to achieve this objective through customer satisfaction. Specifically, they have developed a three-pronged customer satisfaction strategy. The first strategic element calls for retaining their current customers by guaranteeing – as far as possible – that they are satisfied and will consider a Toyota for their next car purchase. (Over 90% of Toyota owners will seriously consider a Toyota for their next purchase. This is the benchmark for the automotive industry and significantly higher than their next closest competitor.)

The second strategic element is the generation of positive word-of-mouth recommendations from current owners to potential owners. Clearly, we have to have satisfied customers if we are going to have any positive word-of-mouth recommendations. Toyota understands the value of both, as well as the customer satisfaction formulae we discussed earlier in this chapter. These demonstrate that each dissatisfied customer we actually hear from in reality represents 36 dissatisifed customers we have not, and will never, hear from.

The third strategic element is the 'conquest' (Toyota's word) of new customers through marketing and advertising to our target audience based upon a clear understanding of the needs and expectations of this audience. Our advertising message can be more efficient and effective if we have a ranked and weighted knowledge of our customers' needs and expectations. An example of this focused advertising message is the 1993–1994 Federal Express television commercials emphasizing Federal Express' ability to provide 'real time' package traceability. As we will see in Chapter 5, Federal Express is emphasizing this benefit because they have ranked and weighted customer data that clearly demonstrate that the traceability of shipments is a major customer 'hot button'.

Do you believe the American and European automotive industries had customer satisfaction as a strategic focus to this degree during the 1970s and 1980s? If they did, why did many American states enact 'Lemon Laws' to protect dissatisfied automobile customers?

STRUCTURE ORGANIZATION FOR CUSTOMER RESPONSIVENESS

The organization that has *Making Customer Satisfaction Happen* as its strategic focus will also have to structure itself accordingly. This organization must be structured to not only serve the front line,

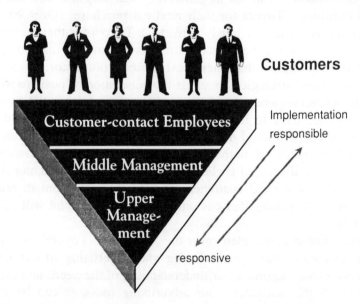

Figure 2.12 Organization structure: the new pyramid paradigm.

customer contact employee in their 'moments of truth,' but also to assimilate customer information and act upon it quickly. It must be an organization structured to act, not react or merely process paper (Figure 2.12).

Therefore, the traditional organization that is 'responsive up' to management must be changed. The *Making Customer Satisfaction Happen* organization has the customer as the focal point of the organization. This approach inverts the traditional organizational pyramid. Management's role is to focus on supporting the initiatives of the customer contact personnel and to provide them with the necessary skills and empowerment to take action to delight our customers. Further, management must ensure that company policies and procedures do not inadvertently hinder its employees. Again, Toyota and Toyota-Lexus are the benchmarks for structuring an organization to focus on its customers (Figure 2.13). This example is detailed later in Chapter 5.

The 'Voice of the Customer' enters the Toyota organization from three different sources – three different customer satisfaction surveys, 800 telephone number (free call) contacts, and other customer contact opportunities. This input flows to the Customer Relations Department which proactively shares it with the relevant parts of the organization.

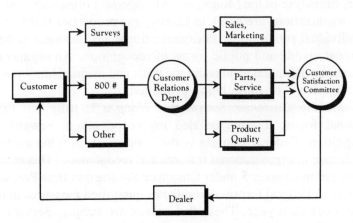

Figure 2.13 Organization structure: the Toyota model.

For example, items relating directly to the car itself (doors hard to close) are forwarded to the Product Quality organization. These groups then take the appropriate actions to address these issues and report their progress to the Customer Satisfaction Committee. This committee is chaired by an executive vice president who reports directly to the president of Toyota.

This is an organization structured with the customer in mind. It is little wonder that the Toyota products are among the most customer-preferred in the increasingly competitive automobile industry.

REWARD AND RECOGNIZE CUSTOMER-FOCUSED ACTIONS

What gets rewarded and recognized gets repeated. It is that simple. Our organization clearly understands where management's priorities are by the actions management takes, and few actions are more obvious than rewards and recognition.

Our management can talk all they want about teamwork, empower-ment, and even customer satisfaction, but if we only reward short-term financial results then that is what the organization will strive to produce, at any cost. Remember that rewards and recognition include promotions, salary increases, bonuses, and special organiza-tional accolades, such as Sailor of the Quarter, Salesperson of the Year, Employee of the Month, etc. Management often fails to realize that organizations are quick to identify the properties that have led to individual promotions, as organizational advancement is one of the most visible and public forms of recognition. An organization will quickly assimilate and adopt the traits and mannerisms – positive and negative – of those being promoted.

If *Making Customer Satisfaction Happen* is truly our organ-izational focus, it must be tied to compensation, reward, and recognition. Federal Express is the benchmark in tying customer satisfaction to organizational rewards and recognition. (This example is detailed in Chapter 5 under Customer Satisfaction Best Practices.)

Briefly, Federal Express establishes quantified measures in three basic areas each year. These three areas are People, Service, and Profits (Figure 2.14).

56

'What gets rewarded and recognized gets repeated.'

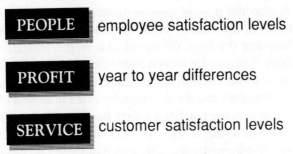

Figure 2.14 Reward and recognition: the Federal Express model.

The People measures involve personal and management objectives. Their successful achievement is based largely on an annual employee satisfaction measure – customer satisfaction of the 'internal customer'.

The Profit measure is based on increases on the previous year's financial results.

The Service measure is based on the organization's performance in satisfying its external customers. The measurement is a compilation of the daily customer satisfaction results that Federal Express senior management reviews each morning.

Failure to meet the agreed-upon measures in *any* of the three categories means no one in the organization gets a bonus – including the chairman! Doesn't that send an unequivocal message to the organization about the primacy of Customer Satisfaction and their role in making it happen?

MANAGEMENT CONTACT WITH CUSTOMERS

Clearly, if our management is to 'own' customer satisfaction, it must 'own' the customer, or at least 'own' a complete and detailed understanding of the customer. Yet, it is rare in most organizations for the management to ever come in contact with the customer. That job is invariably left to the least experienced and least respected group within the organization. Tom Peters has made the point that invariably the person in any organization that has the greatest degree of customer contact is also the person least respected by that organization.

In education, the teacher is the front-line contact with students, yet teachers are the least respected link in the educational chain. School districts across the United States employ high-salaried administrators who wade through the bureaucratic paperwork created by other administrators and do so at salaries often in excess of $100 000 per year. The average teacher's salary has improved markedly over the past ten years, but it still averages about one third that of our paper-pushing administrator. Yet, which of these two has the greater impact on our children, on their future, and therefore on the future of our nation and our way of life? Wouldn't we be better off with two more teachers and one less administrator for that $100 000?

In health care, nurses, house-keeping, and food services employees are in the most frequent contact with anyone staying in a hospital, yet these three groups are invariably on the lowest rungs of hospital seniority and respect. Hospitals pride themselves on the capabilities of their medical staff, yet house-keeping and food service employees have far greater contact with patients than any physician, as we will see in Chapter 3. Yet, how are these critical house-keeping and food service employees perceived by hospital administrations? Are they seen as a crucial customer contact point, or as a large, nameless, faceless minimum wage work force to whom we will turn first if we need to cut budgets within the hospital.

Our field sales people represent our organization to our customers, yet the field sales force is often forgotten or, at best, an afterthought in most organizations.

Consider the situation where a senior executive in one organization was invited to make a presentation on his technical speciality to a customer's management team. The executive was reminded to notify the sales director who was responsible for this customer so that she might attend the presentation and know exactly what was being shared with her customer. The executive curtly replied that he did not want any 'reps' interfering with his presentation! Yet this 'rep' was responsible for this customer that on an annual basis represented over $100 000 in sales to her organization. Shouldn't any right-thinking organization be doing everything possible to support this individual as the interface with our customer? The fact is that the front-line field sales personnel actually bring in the money to an organization and the 'home-office' types just spend it. So much for teamwork and respect for our front-line customer contacts!

Clearly, it is only when our management comes face to face with our customers that they can truly understand their needs and expectations and the priority our customers place on each. Politicians in the United States who are working in Washington, DC like to remain 'inside the beltway' – the circular super highway that surrounds Washington – so that they will not have to face the angry voters who have to pay for the latest folly legislated by Congress and the President. When politicians come face to face with these voters they immediately have second thoughts about their latest budget-busting 'brainstorm'.

One of the foremost executives in consumer goods marketing put the entire customer contact issue into perspective with a uniquely disturbing, but directly applicable analogy. 'The difference between how management currently addresses the customer and how they should deal with them is the difference between dropping bombs from a B-52 at 30 000 feet and shooting someone in the face!'

At 30 000 feet, the human target is nameless, faceless – for all practical purposes they do not exist. That is what our customers are to our organization and to most of management. The customer is a faceless 'them' who no senior management ever sees and who they only hear about through reports which completely depersonalize the customer.

However, when you shoot someone in the face, it is quite a different matter. Someone who could easily pull the lever that drops tons of bombs on people from 30 000 feet becomes quite a different individual when the target is one person, directly in front of you, with a face and a name. When our organization is confronted by an individual customer, we become an entirely different company. In fact, we should see and treat all our customers as if we were face to face with our only customer.

That kind of refreshing thinking should make any management and any organization, regardless of industry or occupation, look at their customers in a new light and realize the tremendous value inherent in each customer.

SUMMARY LESSONS LEARNED

1. Customer satisfaction has hard, quantifiable bottom-line impact.
2. Management must own our commitment to customer satisfaction as the organization's strategic direction.
3. Our entire organization impacts customers and therefore has a role in our commitment to customer satisfaction.
4. Our organization needs to be structured for the benefit of our customers, rather than for the benefit of our organization.
5. Customer satisfaction needs to be proactive, because monitoring complaints and other reactive 'customer service' approaches are inadequate in today's customer-focused, competitive environment.

REFERENCES

[1] Standard & Poor's Stock Market Encyclopedia, November 1993, 15(4), pp. 532, 917, 978.
[2] Buzzell, Robert D. and Gale, Bradley T. (1987) *The PIMS Principles: Linking Strategy To Performance*, The Free Press, New York.
[3] Barker, Joel A. (1992) *Future Edge: Discovering The New Paradigms Of Success*, W. Morrow, New York.
[4] *New York Times*, 19 August 1993, p. 10.
[5] Reichheld, Frederick F. and Sasser, W. Earl Jr., Zero Defections: Quality Comes to Services, *Harvard Business Review*, September–October 1990, pp. 105–111.

Customers and our organization: blessing or burden?

3

> If we are truly dedicated to orienting our company toward each customer's individual needs, then we cannot rely on rule books and instructions from distant corporate offices.
>
> We have to place responsibility for ideas, decisions, and actions with the people who are SAS during those moments of truth – our frontline employees.
>
> *Jan Carlzon*

If we really want our organization to prosper by meeting and exceeding the needs and expectations of both our current customers and those who could become our customers, then we need to understand how our organization appears to these customers. Specifically, we need to understand where, when, and how customers interact with our organization.

In Chapter 1, we defined 'customer' as 'the recipient of our products and services'. It is important to determine just where our organization comes into contact with our customers. This is vitally important because every customer contact, with any part of the organization, is an opportunity to make a favorable impression on our customers.

Traditionally, we believed that only a few parts of our organization actually 'touch' the customer. Most often we think of sales, customer service, and possibly marketing. In fact, our organization is making hundreds of customer contacts each day and each contact

generates a customer perception of our organization. These perceptions accumulate over time and play a dramatic role in determining
whether we retain 'delighted customers' or lose customers who find
us too difficult to deal with.

The examples at the beginning of Chapter 1 are typical of the
contacts organizations have with customers. Yet, management often
loses sight of the 'pedestrian' level at which a vast majority of these
contacts occur. Invariably, these contacts occur in a service environment, where our organization is providing a service to the customer
– whether we believe we are in a 'service' business or not. And
we learned in Chapter 2 that disappointing service is the predominant
reason why organizations lose customers.

You can start a rousing argument with so-called 'product' company
advocates by stating that, in fact, they are actually in a 'service'
business. For some reason, 'service' is viewed with an attitude
approaching disdain by many organizations and their executives.
This attitude probably goes a long way in explaining the remark-
ably poor service levels we experience on almost a daily basis
in our personal lives. The results of this service deprecating
attitude are most graphically reflected in the fact that only three
of the 17 winners of the Malcolm Baldrige National Quality
Award have been service companies – Federal Express (1990)
and AT&T Universal Card Services and the Ritz-Carlton Hotel
Company (1992).

The contention of the 'product' champion in these 'product versus
service' arguments invariably is that the customer is *only* interested
in the product. All the 'service' elements of the transaction –
ordering, shipping, invoicing, and delivery – are viewed by the
'product' champion as tangential to the actual product transaction
and therefore inconsequential. Clearly, the research data presented
in Chapter 2 concerning the primacy of customer service to customer
retention refutes this hypothesis.

Moreover, the research findings are reinforced by our own
personal experiences when we interface with 'product' organizations.
If the bill is incorrect, if the product is late arriving, if what is
delivered is not what we have ordered, do we dismiss these difficulties
as unimportant 'service' issues? Doesn't an organization's failure
in these instances in fact overshadow whatever benefits we may have

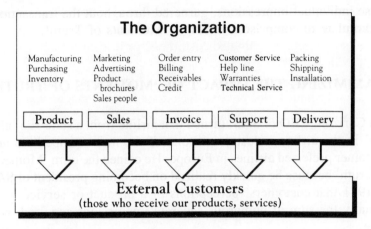

Figure 3.1 Products, services, customers.

attributed to its products? And doesn't every organization have some strong service factors that are an integral part of its organizational makeup?

World Class Customer Satisfaction organizations, such as catalogue shopping giants L.L. Bean and Land's End, realize that the service function is equally as important as the product itself. In fact, the types of products both these organizations sell can be found in a variety of different competitive retail outlets. These retail outlets provide customers with the sensory benefits of actually being able to see, touch, and try on the merchandise (Figure 3.1).

This could be an insurmountable advantage in the clothing business. However, Bean and Land's End compensate for the absence of these sensory benefits in the catalogue shopping mode by providing World Class levels of customer service and customer satisfaction. The telephone purchasing process with either organization is so pleasant and non-threatening because they do not provide the customer with the opportunity or an excuse not to buy.

Bean and Land's End make certain that every customer contact with their organizations reinforces their customer-focused commitment. The entire transaction is viewed as a whole, not as individual sub-elements, because the entire transaction determines what impression the customer develops about the organization and these individual impressions, garnered throughout the transaction, accumulate to comprise a series of 'Moments of Truth'.

MAXIMIZING THE IMPACT OF 'MOMENTS OF TRUTH'

Jan Carlzon, CEO of SAS, turned that airline from a state-run 'also ran' in the airline industry into one of the most competitive and customer preferred airlines in Europe. He coined the term 'Moments of Truth' because he quickly realized on becoming president of SAS in 1981 that customers' perceptions of a product or service were actually the sum of numerous separate impressions acquired over time.

In each of these 'Moments of Truth', customers develop an individual impression of our products, our services, our entire organization (Figure 3.2).

Take a moment to think about all the opportunities you had to form your own perceptions about the airline you flew on your last trip. Think about each of the various 'Moments of Truth' in that

Moment of Truth: An opportunity for a customer to make a judgement about the product or service of your organization.

Figure 3.2 'Moments of Truth'.

experience when that airline had your complete attention and had an opportunity to make either a favorable or unfavorable impression.

As an exercise to demonstrate just how many Moments of Truth can occur for each customer, write down on a piece of paper the Moments of Truth in your most recent airline experience. Then determine what grade you would give the airline for each of these customer contacts. To facilitate your thought process, here are some typical examples of Moments of Truth from an airline flight:

- initial and subsequent phone reservation contacts,
- skycap or curbside baggage check personnel,
- ticketing or re-scheduling personnel,
- speed of entire check-in procedure,
- attitude, personality of check-in personnel,
- cleanliness of terminal and specific waiting area,
- available seating options on the plane,
- boarding procedures,
- gate personnel assisting in boarding,
- departure time,
- cleanliness of the airplane interior,
- flight attendant greeting and attitude,
- promptness of food and beverage service,
- quality of food and beverages,
- flight crew contact,
- arrival time,
- promptness of baggage arrival.

We can see from this simple exercise that on one flight the airline conservatively has had 17 opportunities to make a favorable impression upon us. Similarly, we have had extensive interactions upon which to base our perception of them. The impact of these Moments of Truth cannot be understated. The story is told of some frequent airline travelers who stopped flying a certain airline because the seatback food-beverage trays were always dirty. These passengers made the assumption that, based upon the unkempt nature of these seatback trays, the airline probably did not do a good job on

65

maintenance in total and this maintenance laxity probably extended to the more important mechanical airplane maintenance as well.

This perception may have been incorrect, but the customer's perception is all that counts. They are buying the product and/or service and therefore we – as the provider of that product or service – had better know exactly what is important to them. As we learned in our discussion of the lifetime value of a customer in Chapter 2, our organization just cannot afford to let our key franchise customers – which frequent business fliers are to the airline industry – walk out the door and over to another airline without knowing exactly why they are doing so. We have invested too much time and money in obtaining and retaining their business to willingly let them 'leave the nest' without aggressively trying to keep them.

Consider the number of contacts our organization has each day with its customers. Each of these contacts is a 'report card' for our organization. When we delight customers, we get As. When we satisfy them, we get Bs and Cs. Dissatisfied customers will give us a few Ds and Fs. Do we really know what kind of grades our customers are giving us each day? If we want to know how our children are progressing in school, we review their daily homework, or at least their quarterly or trimester report cards. Wouldn't we want to have the same level of understanding about how our organization is progressing in the marketplace?

In Chapter 2 we talked about a 'macro' approach to customer categorization when we discussed the three types of customers. We need that broadbased information about our customer categories – dissatisfied, satisfied, delighted – but we also need to know the same information about each contact our customers have with our organization. This is the 'micro' approach. An important first step is to have the organization and its management realize that these contacts – these Moments of Truth – do occur and, in fact, are occurring as we read this.

A first step in dramatizing this to an organization is to have the organization's management board – or Quality Implementation Team – conduct a Moments of Truth exercise, similar to our airline one, for their organization. We can start by brainstorming the opportunities our customers have for interacting with our organization. Then we can start to make some rough estimates concerning

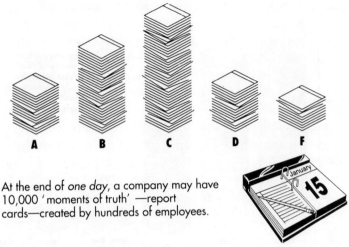

At the end of *one day*, a company may have 10,000 'moments of truth' —report cards—created by hundreds of employees.

Figure 3.3 'Moments of Truth': one day's results.

the number of these contacts per week, per month, per year (Figure 3.3).

Invariably, the results come as a great surprise to almost every organization. Often we are unaware of all the customer contact points within our organization and we are many times disconcerted by our lack of knowledge concerning how we are perceived in these interactions.

One organization actually conducted detailed research in which they identified and quantified their Moments of Truth. This company enumerated over 1 500 000 customer contacts on an annual basis. Interestingly, the vast majority of these contacts came in the order entry process, not in the areas we traditionally think of as critical customer contact points.

In the case of our organization, we have to ask if our employees, who are interacting with our customers on such a frequent basis, are trained for this interaction (Figure 3.4). The organization conducting this research realized, as a result of this study, that its employees in customer relations and the sales organizations had some degree of training in customer interaction, but even that training was limited. What training the customer relations and sales personnel had

67

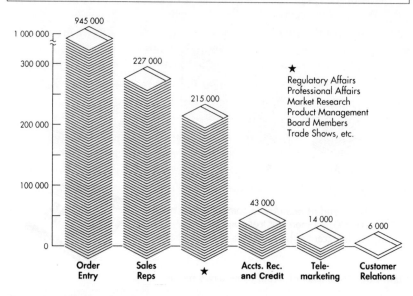

Figure 3.4 'Moments of Truth': one company's experience.

received was invariably focused on communicating one way – communicating the organization's desired message to the customer. These employees were woefully unprepared to serve as receptors of customer communications. All the other employees who were now known to interact with our customers had received no special training, not even concerning basic phone etiquette.

Therefore, a clear first action step, deriving from this basic 'organizational audit' on customer interaction, was the development of a training program for all employees who came into any contact with the customer. This training program featured a basic 'core' program applicable to all customer contact employees. An integral part of this training was an emphasis on 'active listening' in order to capture key insights provided by the customer. Additionally, customized modules were developed for those with particular and specialized job functions interacting with customers. Included in this group were the sales and tele-marketing groups who required specialized 'selling skills' training.

Once an organization determines how often it is actually interacting with its customers and who in the organization is doing this

68

interacting, we also need to determine what impression we are making in each contact, what information on what subjects is being discussed, and what we are learning from each of these contacts. Also, does our organization have a mechanism for sharing this knowledge from customer contacts with all parts of our organization that can really do something about it?

For example, the organization that conducted the quantified Moments of Truth research determined that a great deal of organization commentary on a variety of subjects was occurring during the order entry process. The customer was making comments about the ease of ordering, shipping instructions, satisfaction with previous orders, questions on current and future pricing, and overall ease of dealing with the organization.

The order entry personnel were not trained on how to receive this information, nor were they equipped to capture this information efficiently or effectively. Some would jot notes about specific conversations. However, since the Order Entry department had its productivity measured upon the number of calls handled, it was not financially in the best interest of these employees to take the time to document customer comments.

Clearly, this valuable customer information was also not being shared with the other parts of the organization that would benefit from receiving it. For example, package engineering should want to know that the product's shipping containers were easily dented. Finance might want to know that payments were delayed because invoices were complicated and difficult to read. Operations should want to hear that our organization's order fill rates were well below those of our competitors, at least for this customer.

This type of information is invaluable to an organization because it enables the separate functions within the organization to calibrate their internal standards based upon the actual needs and expectations of the customer. We will discuss this calibration process in further depth in Chapter 4.

The organization then developed action step two, which followed the training of all customer contact personnel. In this step they instituted a process for collecting customer comments and also changed the incentive process for the order entry department to more clearly reflect the customer's needs, rather than merely the organization's.

Their new measures included Order Accuracy and Customer Satisfaction with the order entry process, as reflected in annual Customer Satisfaction studies (more on these in Chapter 6).

Once we know where customer contact is occurring within our organization, who is being contacted from our organization, and what information is being exchanged, we then need to determine how our customers perceive our organization's performance during each of these Moments of Truth. We need our organizational report card. We need to know how many As and Bs we are getting and where the Ds and Fs are. This will enable us to prioritize our corrective and preventative efforts and really enable our organization to take a proactive approach to Customer Satisfaction. Does our organization currently have this level of knowledge? The World Class Customer Satisfaction organizations do!

Concerning where customers are interacting with an organization and what information is being exchanged, Federal Express knows on a real time basis the exact status of every package shipped with them – over a million each day. They know if they have arrived on time, how many were late, where there were billing disputes, where there were satisfied and 'delighted' customers and where there were unhappy customers.

The Federal Express management board reviews these results daily and the company's performance on key criteria related to these daily reports is a contributing factor concerning whether anyone in the organization gets an annual bonus or not. Importantly, holding compensation hostage to quantifiable Customer Satisfaction results has the benefit of galvanizing the organization's attention to this issue.

On the issue of providing customer contact personnel with the appropriate level of training, Federal Express provides all customer service agents – the customer's phone contact to Federal Express – with five weeks of training before they are allowed to handle calls on their own. They receive another four hours of training each month, and twice each year have to be recertified for their job by taking a test via computer. This guarantees that the Federal Express employees that are interacting with customers are 'on top of their game' at all times and stay that way [1].

So, not only can Federal Express tell their customers where their packages are at any one of 20 separate parts along their journey,

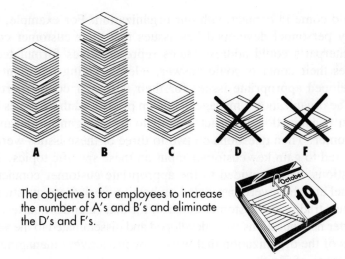

Figure 3.5 Managing 'Moments of Truth'.

they also have well-trained and prepared individuals at the other end of the phone when the customer calls on any matter. That is managing the 'Moments of Truth' (Figure 3.5). (A further discussion of Federal Express' 'Best Practices' concerning the empowerment of its 'frontline' customer contact personnel is included in Chapter 5.)

Another important aspect of managing the organization's Moments of Truth is moving these situations from reactive to proactive ones. A major organization in the vision care industry realized that this multitude of Moments of Truth actually represented an unprecedented opportunity to proactively obtain additional information about their customers. They further realized that they could actually obtain this invaluable customer information for little or no incremental cost. This was seen as a 'win-win' process, since the customer had initiated the call in most instances and the value of the information received far outweighed the minor amount of incremental time their customer contact personnel might remain on the phone or otherwise in contact with the customer.

Therefore, the organization developed a listing of key issues on which they wanted further customer input. These issues were differentiated by the segment of the customer's organization that would come in contact with our organization. For example, order entry personnel developed key issues that their customer contact counterparts could address, sales representatives formulated key issues their contacts could answer, tele-marketing representatives developed appropriate issues for their customer counterparts.

The organization's management then prioritized these issues based upon organizational impact – both to their organization and the customer. Then questions on two to three of these issues were formulated to gain key customer input on these specific topics. These questions were provided to the appropriate customer contact personnel and they were instructed on how to work these questions into their traditional customer contacts for that calender quarter. Each quarter new questions were developed and disseminated to the various parts of the organization that were now proactively managing their Moments of Truth.

The organization soon found that it was obtaining exceptionally useful customer input on a variety of subjects. This information augmented and complemented the organization's existing market research efforts. In fact, the two research areas worked in concert. The traditional market research studies would enumerate customers' prioritized issues and questions on these issues would then be formulated for use in the customer contact research.

Initially, the market research professionals were somewhat skeptical of this 'non-traditional' research approach. However, considering that the organization was 'touching' its customers over one million times a year, information received with this level of respondents appeared to be just as reliable as structured market research. This 'traditional' research might be conducted with several hundred, or in the case of major 'habits and practices' research, a few thousand respondents. The 'non-traditional' research was routinely tapping into several thousand respondents and therefore represented real value, even if the questions might not be structured to satisfy a research purist and despite the fact that the the order entry personnel were not classically trained telephone interviewers.

Clearly, if a few hundred of our customers are saying that our invoices are difficult to decipher, we probably do not want to spend the time doing a statistical test on the 'significance' of this response rate. Rather, we had better correct our invoice format before our competitors further capitalize on our failings.

This approach is consistent with our basic Quality approach of being proactive, customer-focused, and managing our organization to exceed our customer's needs and expectations. We should now realize that these customer Moments of Truth can be used to our advantage to achieve our objective of retaining our current customers and attracting new ones to our franchise.

'MOMENTS OF TRUTH' MINI-TEST

Here is a quick exercise we can conduct to determine how our organization may be faring during these critical Moments of Truth. Clearly, it is not a quantifiable study and not one which is totally projectable to our entire organization. However, we can probably obtain a relatively accurate customer perception of our organization from some, or all, of these mini-test actions.

1. Call the main telephone number of our organization during normal work hours. How many times does the phone ring before it is answered? What is the initial response and tone of the person answering the phone?
2. Call our organization with a basic question concerning any of the 'core capabilities' of the organization. How long does it take to obtain an accurate answer and how many transfers were required before the question is answered completely?

     For example, call a bank and ask about their 30-year mortgage loan rates for a new home purchase. Call a school and ask the date of their spring semester vacation or the school's average score on a standardized test, such as the Scholastic Aptitude Test (SAT) which is integral to most college admissions processes. Call a consumer products company and ask them what sizes their leading product is sold in and the name and address of the store closest to you that carries this product.
3. Call our organization during 'off peak' hours, as if we were a customer calling from a different time zone – somebody from California calling New York, somebody from Moscow calling

London, etc. Are we able to reach anyone of authority and could we transact any business with this person if we had wanted to?

4. Visit the neighborhood around our facility and ask residents their impression of our organization, as if we were writing an investigative newspaper article about the organization. How would they describe the organization in five words or less? Could they describe the organization and its character at all?

5. Would we recognize the most senior management personnel at our top three customers if we passed them on the street? Can we name and give the direct office telephone number of our organization's key contact at our three largest customers? Can each of the members of our management team do likewise? Do our top three customers know our home phone numbers and have they ever called us at home just to 'chat'?

There is no need to score this exercise. The results we achieve should make it fairly obvious what actions need to be taken. Importantly, this exercise begins the process of getting our organization to begin looking at itself from the customer's point of view, rather than from the perspective of 'we've always done it this way and customers will have to fit our needs' that occurs in so many organizations.

SUMMARY LESSONS LEARNED

1. Our customer satisfaction efforts must be focused on our external customer first. Our policies and practices impacting this external customer can be applied to internal customer–supplier relationships, but our primary focus must be on the external customer.

2. 'Moments of Truth' occur between our organization and customers and they directly impact our customer's perceptions of our organization, our products, and services.

3. Our organization needs to determine where 'Moments of Truth' occur in our organization, how often they are occurring, and what impressions are being generated by our organization to customers during these interactions.

4. We can proactively use these 'Moments of Truth' to market our organization's products and services to customers if we know who is 'touching' our organization, where these contacts occur, and what information is being shared during these interfaces.

REFERENCE

[1] *Blueprints for Service Quality – The Federal Express Approach*, American Management Association, New York, (1991), p. 20.

The *Making Customer Satisfaction Happen* model

4

To satisfy the customer is the mission and purpose of every business.

Peter Drucker

Before we can have results similar to those of Federal Express or any other World Class Customer Satisfaction organization, we need to develop an overall approach – or process – for *Making Customer Satisfaction Happen*.

Our recommended approach is a very simple, yet comprehensive model. It has been used in a wide variety of organizations, because while specific customers may vary, the overall approach any organization should use to address these customers need not vary. Throughout our discussion of this model, we will provide examples of its use in a variety of organizations, from large and small companies to non-profit, educational, and even government organizations.

The four-part Customer Satisfaction Model presented is drawn as a circular, continuous process. The model is purposely designed in this manner to reinforce the continuous nature of our Customer Satisfaction efforts and to remind us that we continually need to revisit each segment of the model on a regular basis. Our approach to Customer Satisfaction is not a static, 'one time' event. Rather, it represents our organization's strategic approach and therefore a constant in our operational equation for success.

We will briefly overview our Customer Satisfaction Model now and then detail each segment separately (Figure 4.1).

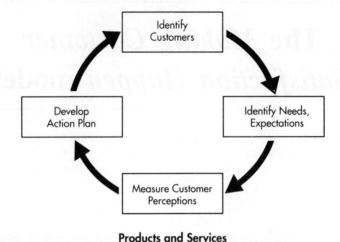

Products and Services

Figure 4.1 The 'Making Customer Satisfaction Happen' model.

STEP ONE – IDENTIFY CUSTOMERS

The first step in our model is to *identify our customers*. We need to know who they are currently and who they could be. We need to understand our 'chain of customers' – such as distributors, wholesalers, the retail trade, and others who form a chain to our final, end customer. We will also have to prioritize the importance of our various customers, because we will not be able to address all of them concurrently. That is one reason why our model is circular – it is a continuous process, not a one-time event.

Many organizations become confused at the outset of a discussion on Customer Satisfaction. They believe that every organization has *one* customer, and once they find that customer and satisfy them, their task is complete. This theory is flawed on several obvious levels, yet it is a surprisingly widely held approach to Customer Satisfaction and the identification of an organization's customers.

For example, many hospitals identified physicians as their primary customer when asked to name and prioritize their organization's customers. If the physician is the hospital's sole, or even primary customer, then who is that person whose gall bladder we just

78

removed, who are those people visiting the recovering patient, what are the various organizations paying for this procedure, and what are the regulatory agencies overseeing our entire medical structure?

Or consider the major consumer goods firm. They like to focus on the 'consumer', not the 'customer'. The Consumer is seen as the product's 'end-user' – the person who drinks the orange juice, washes the clothes, diapers the baby. The Customer is identified by this organization as the retail and wholesale 'trade' who serve as the conduit through which manufacturers are regrettably and unfortunately, in their view, required to pass on their distribution path to the Consumer.

Clearly, this is merely a semantics game. The 'trade' is a very important customer, as Procter & Gamble learned when they tried to bully WalMart Stores' Sam Walton, the richest man in North America at the time and the owner of one of the most powerful mass merchandiser chains in the world.

Procter & Gamble had built its phenomenal consumer products growth on going direct to the 'consumer' through television and print advertising and promotions, such as price-off coupons and pre-priced products. They viewed the grocery and drug 'trade' as a necessary evil and treated them accordingly. Historically, Procter & Gamble had not been adverse to pulling some of their high volume, traffic generating products, such as Tide detergent, Crest toothpaste, or Folgers Coffee, from accounts that balked at buying recommended quantities of new product introductions. This tactic failed miserably when faced with the tremendous retail impact of the WalMart stores. Through their volume and impact within the communities in which they were located, these stores could really dictate to manufacturers, rather than vice versa.

To Procter & Gamble's credit, they quickly realized this fact when they were unable to 'muscle' WalMart. Completely reversing their position, Procter & Gamble realized WalMart was a 'customer' and then went about setting up offices next to WalMart headquarters in Arkansas to make certain that Procter & Gamble clearly understood, met, and hopefully exceeded each of this newly found customer's needs and expectations. Not all consumer goods organizations have been as quick to 'unlearn' their previously flawed thinking about exactly who is and is not a customer. Such organizations are at a distinct disadvantage when faced with a competitor who is not only 'in touch' with the customer, but has actually built offices next door!

In another discipline, many educational institutions are equally puzzled in pursuit of their organization's 'one' customer. Is it the school board, the school's administration, the faculty, the parents, the college or university board of trustees, or the students? Maybe it is the community served by the educational institution or possibly the taxpayers who are funding the public or state-operated schools. In fact, we would be hardpressed to find anyone in education who views the taxpaying public as a customer of any nature.

An excellent example of this disparaging attitude toward the taxpaying public on the part of the educational hierarchy was demonstrated when a local school board's budget was rejected by the taxpayers in that community. The incensed school administration – staffed by a full complement of highly paid administrators, advisors, consultants, and a whole host of other titles who had never set foot in an actual classroom – publicly stated that the voting public should not have the right to vote on their budget because the public was just not sophisticated enough to understand it!

If the taxpaying public that voted down the school board's budget were comprised of graduates of the school system we are discussing, there was a good chance they probably would not have even been able to read it, let alone understand it, with all the money that was being budgeted for non-classroom bureaucrats! Yet haven't we previously defined 'customers' in Chapter 1 as the recipients of our products and services, and by extension, the person paying for them?

In fact, organizations have multiple customers. Do not be confused by the semantic puzzle of our consumer goods organization – they are all customers. The sooner we recognize that fact, the sooner we can be about the work of identifying all of them, prioritizing their impact and importance to our organization and then learning everything there is to know about them.

STEP TWO – IDENTIFY CUSTOMER NEEDS AND EXPECTATIONS

The second step in our Customer Satisfaction Model is to *identify customer's needs and expectations*. This identification of needs and expectations will come from the customer's vantage point, not ours. We want them to tell us what *they* want and expect – what is important to *them*, what is a 'need-to-have' and what is a 'nice-to-have'. As part of this identification we will also learn how our customers prioritize their needs and expectations – what is *most* important, what is less important.

This prioritization of our customer's needs and expectations is critically important. World Class Customer Satisfaction organizations can literally rank and weight their various customer's needs and expectations. They not only know what customers believe is most important, but also what emphasis customers place on these various factors. The organization armed with this type of customer knowledge can therefore focus on meeting and exceeding these prioritized customer needs and expectations. There are no unnecessary steps, no misdirected actions. They are efficiently and effectively focused. This is not only a tremendous advantage in their customer relationship, but also versus their competition.

Consider the case of the large health care product distributor that was developing a major service offering for its customers. This distribution organization was considering the development of a Supplier Quality Management module that it would provide to its hospital customers as a value-added service. Clearly, there is a need for this service, as hospitals are faced with a wide array of suppliers for a variety of products and services. Moreover, hospital personnel responsible for this supplier interface have often not had the benefit of extensive training on the subject of supplier management that some manufacturing firms, such as a Toyota, Ford, or General Motors have had.

A consultant was called in to assist in the development of this Supplier Quality Management program. However, her first step was a very 'unconsultantish' one. She asked her client if they knew the customer's actual priority on the subject of Supplier Quality Management. Specifically, where did this subject rank in the customer's hierarchy of informational needs? The answer came back that the organization had conducted research with a variety of hospital customers and had asked each one if they would be interested in a program on Supplier Quality Management. In each case the reply was positive. This represented sufficient data for the distributor organization's management to proceed with the project and retain this 'doubting Thomas' consultant!

She persisted in her questioning of the client. Yes, the hospitals had said Supplier Quality Management was important, but in relation to what? How did it rank in comparison to waste treatment and removal, hazardous 'red bag' disposal, patient-focused care, managed care marketing implications, capacity utilization maximization, etc. On a comparable personal life basis, the question had been asked, 'Do you like vanilla ice cream?' The reply was 'yes'. However, that affirmative response did not signify that vanilla was, in fact, the customer's favorite, only that it was acceptable in the absence of other alternatives.

In fact, vanilla may well have ranked number ten on a list of favorite flavors, but we will never know unless we ask the customer's preferences in relation to a series of options.

Therefore, the organization could have made a substantial investment in time, material, and consulting fees and developed an outstanding Supplier Quality Management presentation that was relatively unimportant to customers when confronted by competitive proposals that addressed their top three rated issues. So prioritizing our customer's needs and expectations enables our organization to focus and focus is critical to efficient and effective Customer Satisfaction.

STEP THREE – MEASURE CUSTOMER PERCEPTIONS

Armed with the information on what our customers believe is important, we can now ask them how they perceive we are doing versus those expectations. This is the Measurement of Customer Perceptions. We can also ask them to compare their perceptions of our performance with that of our competitors.

Once we know how we are perceived by customers and how they perceive our competitors, we can develop action plans to address areas where there are important gaps between items the customer has identified as a priority and where they perceive our performance is trailing behind that of our competition.

This is an ongoing process of continuous improvement, hence the model's circular design. We will continually need to determine who our customers are – who comprises the total market for our products and services. Customers' needs and expectations are continually changing. The 'bar' is constantly being raised. Therefore, we will continually have to monitor their needs and expectations for changes, additions, deletions, reprioritizations.

We will need to measure customers' perceptions of our performance as frequently as possible and feasible. Clearly, the more current this information is, the better able we are to be proactive rather than reactive. The era of managing a consumer products business through bimonthly A.C. Nielsen data with a two-month lag time from survey to availability is over and probably ended 20 years ago with the demise of '1960s Marketing'!

Fifteen years ago even small, but progressive food chains had access to daily computer reports of in-store sales volume. They could make intelligent, data-based decisions on this information. The people presenting products to them in the selling situation – even from the nation's largest product firms – were presenting sales information based upon two month-old Nielsen projections, not even the actual sales data. With the enormous number of customer contacts our organization makes on a daily basis, as discussed in Chapter 3, it is unimaginable that our organization cannot have a weekly, if not daily, view of our customer's needs and expectations and our relative performance versus that of our competition.

STEP FOUR – DEVELOP ACTION PLANS

Our action plans will be continually refocusing on the issues that impact our customers today. This action plan for addressing Customer Satisfaction issues will translate directly into our overall organizational strategic plan.

Finally, we need to understand that our customer is looking to us not only for products but also for the services which accompany those products. We may not think of ourselves as providing 'services', but the customer sees invoicing, accounts payable, delivery and customer service as *services* that we provide. If the customer sees them that way, we had better see them that way as well and we had better know how we are perceived by our customer in performing each of these. This information will have a direct impact on developing our action plans.

STEP ONE DETAILS – IDENTIFY OUR CUSTOMERS

Our first step is to identify who our customers are (Figure 4.2). We spoke earlier about our broad approach to customers – both ours and those of our competitors. Let's refine this slightly by looking at our current customers, our former customers, and those who could be our customers but currently belong to our competitors. (Of course our former customers could be a sub-set of our competitors customers' as well).

We need to identify who our current customers are. Just who is it that comprises our current franchise? Who actually purchases our

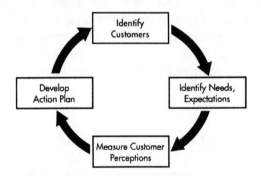

- **Current customers**
  (professionals, end users, purchasing agents, associations, distributors, regulatory agencies)

- **Former customers**
- **Competitors' customers**

Figure 4.2 Step one: identify customers

products and services and who directly influences these purchases? We have to make certain that we have identified all parties that influence the purchase of our product and then understand their relative importance in that purchase decision.

For example, a surgeon may use a particular medical instrument or device. The chief operating room nurse may control what instruments and devices are present for each procedure, and the vice president of materials management may actually sign the purchase order for the device. By now, we should know that these individuals are all customers – that level of knowledge alone puts us in pretty rarified company among our competitors who are targeting only one of these three – but what is their relative level of importance in the buying decision? We can only learn that by working closely with these customers and understanding their unique process. Once we know who the key decision maker is, we can focus our information gathering efforts on them.

This may sound like common sense, and it is. However, major decisions have been made on substantial product lines without a clear understanding of the relative importance of certain decision makers. For example, from our earlier discussion of disposable diapers, we know that many manufacturers in this market made critical marketing decisions based upon their perceptions of the customer's needs and expectations. One manufacturer prided itself on the fact that its diapers were what babies would buy, if they could. But babies don't buy diapers – wine and cigarettes maybe, but not diapers! Their competitor was focusing their marketing efforts on the mother and her needs and expectations. By listening to the actual purchaser of the product – the 'gatekeeper' in the terminology of several major consumer products organizations – this diaper manufacturer was able to 'leapfrog' its competitor producing the 'baby-preferred' diaper and eventually drive them out of the market entirely. So knowing who the priority customer is can be critical.

We can learn who these customers are by talking with our organization to understand the *process* involved in the purchase of our products and services. Who places the orders? Who needs to authorize them? Who recommends? Who can cancel them? We can learn most of this information by studying the actual process and

85

by conducting qualitative research within our organization with those departments and functions that deal with the customer.

As we progress further with this model, more and more information will have to be developed using basic market research. While most major companies have a market research capability, they may not be entirely conversant with the type of research required. Other organizations may have little or no research experience, such as educational institutions and non-profit groups.

It is not always necessary to contract with professional Customer Satisfaction research firms if we are seeking preliminary levels of customer understanding. A good starting point for an organization in this situation is to discuss our customers and their needs and expectations with those members of our organization who most frequently come in contact with these customers. Additionally, informal 'focus group' qualitative discussions with our customers can also help us gain a clearer understanding of their prioritized needs and expectations. As our information needs expand and our customer base size increases, we are well served by dealing with professionals.

Therefore, it is necessary to identify organizations that are more than just market research firms. We need to identify firms that specialize in the area of Customer Satisfaction research. These firms are discussed in detail in Chapter 6.

It is not the objective of *Making Customer Satisfaction Happen* to make us into market research technicians. Therefore, no time or discussion will be spent on the development or execution of research questionnaires. This is always best left to the professionals who we will recommend for your consideration.

Nor should organizations believe they can seriously conduct their own Customer Satisfaction measurement without some professional input. Neither should they believe they can merely tailor someone else's approach to their needs. One organization we dealt with phoned their enthusiastic support upon learning that we were developing an approach for *Making Customer Satisfaction Happen* and asked to be sent a copy of the questionnaire! A frustrated colleague, who is slightly less tolerant of those seeking shortcuts to Customer Satisfaction, suggested sending out one questionnaire and a bottle of 'White Out'. This type of research is just too important for us to adopt a 'one size fits all' approach.

What organizations should do is to learn from the experiences of other organizations who have had extensive experience in customer satisfaction measurement and from the firms that conduct this research. The World Class Customer Satisfaction research organizations are very open in discussing the basic approach they take to measuring customer perceptions of our organizations.

Prototype Customer Satisfaction measurement processes and actual questionnaires are included in Chapter 6. They are included for reference and information only. They are not to be seen as definitive or applicable to all situations for any and all types of organizations. Rather, they have been included as *examples* of what Customer Satisfaction supplier presentations and actual studies *can* look like.

In this first step in our *Making Customer Satisfaction Happen* Model we also need to understand who our former customers are and why they are *former* customers. This will also require some research. Possibly we can meet with these former customers and learn of their concerns/problems with our operation. In other instances we may have to depend on third party research – where we are not identified as the source of the research – to clearly bring out their real concerns.

It is important not to be satisfied with the initial, superficial reasons former customers may give for moving their business from us. We are now going to examine a classic example where banks initially believed customers were switching accounts based on rate differences, when the real reason for this switch was poor service. In fact, we saw that 68% of dissatisfied customers who leave do so because of *service* issues. Therefore, do not assume that *price* is the reason – even if that is given as the initial response. It is *seven times* more likely to be service related (68% to 9%).

The classic study in this area was conducted among bank customers, in particular those closing out their accounts at one bank and moving them to an entirely new financial institution. The bank from which these accounts were departing was sufficiently enlightened to arrange and conduct some 'exit' interviews with these soon to be former customers. When the individuals were queried about why they were leaving Bank 'X' for Bank 'Y', they invariably stated that their new bank offered 'better rates' on savings accounts, loans, and other financial instruments.

Upon receipt of this research, Bank 'X' immediately corralled its bank officers and set the stage for a major review of bank loan and savings rates. Two perceptive bank officers at Bank 'X' interrupted the proceedings by noting that the rates Bank 'Y' was advertising in the newspaper ads were, in fact, less attractive than those already in place at Bank 'X'! Therefore, something appeared to be wrong with the research results or with the actual research instrument.

Qualitative, one-on-one personal interviews were arranged with some of the Bank 'X' customers who had moved their accounts. In the more relaxed environment of a personal interview, the actual reasons for the bank account moves began to surface. In all probability, you can probably identify the actual reasons for these account switches. Among the real issues identified by dissatisfied Bank 'X' former customers were: 'inconvenient lobby hours'; 'crowded lobby and long lines'; 'indifferent tellers and not enough of them at that'; 'turned down for a loan I clearly qualified for'; 'treated like another number, not as a person'.

These are all service issues and these actual results are completely consistent with the relationship between service and price that we saw in Chapter 2 concerning why businesses lose customers. Just as typically, management's first reaction was to focus on price (loan and savings account rates), in part guided by misleading research. They started down the wrong corrective action path because they lacked a clear understanding of their customers, their customers' needs and expectations, and even their organization's relative competitive posture.

Fortunately for Bank 'X', they discovered the error of their ways and began to focus on the service issues that really mattered to their customers. Unfortunately, they had already lost a large number of customers without ever knowing why and without ever endeavoring to get them back by addressing their real concerns. The lesson learned in this example is to never be satisfied with vague, surface solutions to problems or situations, particularly when they involve our valuable customers. And we need to continually remind ourselves that all our customers are valuable. If we do not believe our customers are valuable to our organization, there are many other organizations in the marketplace who do.

We also need to learn about who comprises our competitors' customers. How are they different from ours? If they are made up solely of our former customers, what are our competitors providing that we did not? Far too many organizations are too willing to categorize their competitors' customers as less sophisticated, less discerning, less discriminating, or just plain less intelligent than their own customers.

This is just nonsense. Consider your personal life experience. Why do you switch products and services? Can you think of a recent example where you switched from the product or service which you had previously used? What was the reason for the change? Does your organization offer a similar opportunity for customer service dissatisfaction to your customers?

In that regard, we need to understand the real reasons why our competitors' customers are not ours. Often, the real reasons are too painful for our organization to bear – possibly our products or services are significantly premium priced but with no discernable product or service differentiation versus competitor.

Management invariably does not want to hear that type of information. They attribute all manner of customer benefits and special properties to our products and services which the customer either cannot differentiate, or which are of no real importance to the customer. That is why it is critical to determine what the customer's needs and expectations are, not what we *perceive* them to be. We need to enter into this journey with the belief that our competitors' customers are just like ours and could easily be ours. It is our job to determine why they are not ours. Remember, it is probably *not* price alone, as Bank 'X' painfully learned.

Of course one of the best ways to understand both our competitors and their customers is to 'manage by walking around', as Tom Peters described in his seminal works on Customer Satisfaction, *In Search of Excellence* and *A Passion for Excellence* [1]. One of the best and most noteworthy practitioners of this management style was Sam Walton of Walmart. His was a management style of walking around, and literally flying around, to visit as many of his stores as possible. He continued this practice until bone cancer finally stopped his one man crusade for first hand customer and competitive understanding.

Sam Walton was not satisfied with merely visiting his stores, speaking with his employees, reviewing the financial results of not only each store, but also of each department within each store. He also took the time to visit competitive stores that appeared to be making inroads on WalMart based upon his review of each store's financial statements, for Sam Walton was a manager who embraced our previously discussed motto of 'In God we trust, all others bring data'. Sam Walton walked the aisles of competitive outlets and compared their prices and values with those in his local WalMart. It was not uncommon for the local WalMart store to take immediate and direct merchandising and or pricing actions based upon 'Mr. Sam's' competitive intelligence forays [2].

This is not the type of management style one might expect from someone who at the time was the richest individual in the United States. However, it is exactly this management style, and its direct access to the customer and competition, that uniquely explains Sam Walton's phenomenal success.

As we conclude our discussion of this first step in our Customer Satisfaction Model, this is an appropriate opportunity for us to do some 'hands-on' examination of our own organization and its customers. Therefore, here is a brief exercise that can be an ideal start to our Customer Identification process. The ideal group with which to initially conduct this exercise would be our senior management. Thereafter, it can be an excellent vehicle for stimulating interest in the topic of Customer Satisfaction and the critical need to identify our customers at the outset of that process.

---

1. Brainstorm a list of our business' most valuable customers. Be as specific as possible. Instead of 'distributors,' say 'Joe Smith in Omaha'.

2. Reduce our list to the five to ten most valuable customers. We determine what constitutes 'value' – sales, industry influence, growth potential, etc.

---

3. Put ourselves in our customers' shoes and develop a list of 'Moments of Truth' – opportunities those customers have to judge us. Some sample moments of truth: 'Opens our invoice'; 'Calls office to speak to sales rep'; 'Drives by corporate offices'; 'Sees our product being used by a colleague'. We should be able to list at least a dozen 'Moments of Truth'.

4. Select one 'Moment of Truth'. List those people or functions within our organization who have had some influence over what the customer sees, hears or experiences. For example, if we select 'Opens our invoice', we might list 'software engineer (who designed the form)', 'sales person (who filled out order)', 'order entry clerk', 'purchasing (who bought the paper, envelopes)', 'mail room personnel (who delivered them to the post office on time)', etc. The point is to understand who has an impact – direct or indirect – on how customers judge us. Develop detailed lists for three 'Moments of Truth'.

5. Discuss what would cause a customer to grade us as a 'D' or 'F' during a 'Moment of Truth' and who would be responsible and empowered for immediately improving this grade.

6. Determine actions that would prevent these low scores and decide which of these could be implemented immediately.

Our learning from this exercise should be that there is probably some degree of confusion among the organization concerning who constitutes our key customers. There is probably also an important gap in our knowledge about how often and where our organization interfaces with these customers. Finally, there is probably also an important information gap concerning how our customers view us in each of these critical 'Moments of Truth'. This type of learning should not scare or deter us. Rather, it should invigorate us for the journey to World Class Customer Satisfaction that lies ahead.

STEP ONE SUMMARY ACTIONS

Here are some summary actions to consider as we begin to identify our customers:

1. Who makes up our current universe of customers? Do we deal with a supply chain of customers? If yes, who are they?
2. How would we prioritize our customers? Is it by sheer volume, potential, risk of loss, etc.? Who are the 'influencers' in our customer equation?
3. Who in our organization comes in contact with these customers? Do we know what these key contact personnel have to share about these customers?
4. Who are our former customers and why did they really leave?
5. Who are our competitor's customers and what keeps them from being ours?

STEP TWO DETAILS – DETERMINE CUSTOMER NEEDS AND EXPECTATIONS

The next part of our model involves the determination of our customers' needs and expectations concerning our products and services. It is critically important that these be the customer's needs and expectations. These are *not* to be *our* perceptions of what we *think* our customers need and expect. Rather, we are looking for the voice of the customer to tell us their expectations of our products and services. Far too often, management has believed it has a divinely granted perception concerning the customer's wants and needs. This is the hallmark of '1960s Marketing', as we discussed earlier.

This subject of Determining Customer Needs and Expectations may be a source of controversy and defensive behavior among some in the organization – particularly marketing and sales (Figure 4.3). They have long been viewed by the remainder of the organization as the 'keepers' of customer knowledge. They may, in fact, be the keepers of customer *data*, but they usually lack real customer *knowledge* – particularly hard, quantifiable, ranked and weighted knowledge on customer needs and expectations. When we are dealing with something as critically important to our organization as our customers, it is imperative to manage by facts and not opinions. Remember the motto of 1990s marketing, 'In God we trust, all others bring DATA!'

There are a myriad of examples of organizations incurring extensive and unnecessary expenses because they believed *they*, not

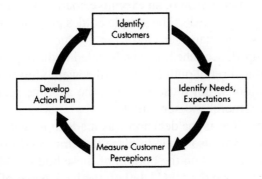

- Determine the characteristics of your product and services (such as size, price, color, availability, support, etc.)
- Ask customers to rank characteristics according to relative importance.
- You now have a list of what customers need and expect.

Figure 4.3 Step two: determine customer needs.

93

the customer, knew best. Here are some that might apply directly to our organization.

Many organizations are very proud of how their actual product looks. The color and clarity of the product many times helps differentiate it from competition, or so management believes. One manufacturer believed this to such a degree that they bottled their product in Lexan plastic bottles. Lexan is a premium plastic with exceptional clarity and strength. It is often used in the production of jet airplane canopies, where clarity and strength are prerequisite. This premium strength and clarity came at a cost, and Lexan was significantly more expensive than the plastics previously used in this product packaging. Management believed that price premium was acceptable given the product's aesthetics and passed that price increase on to the consumer.

We have to remind ourselves that when management was reviewing this proposal, and examining the Lexan bottles versus their poly vinyl chloride (PVC) precursors, the examination was conducted in the pristine environment of an executive boardroom, not the razzle-dazzle of the grocery aisle. The sample bottles presented for management's review were hand selected from production runs, they were not price stamped, scratched, or manhandled to the degree one might find in the typical retail outlet. The bottles were viewed under ideal lighting conditions, not those found in the average retail grocery, drug, or mass merchandiser outlet.

At the same time as the decision was made to go with Lexan bottles, a decision was also made to place the Universal Product Code (UPC) bar code labels on the bottom of these bottles. Management believed that to have the UPC label on the back panel of the beautifully clear plastic bottles would detract from their customer appeal. Never mind that the presence of a front label on these bottles made it almost impossible for consumers to see the UPC back label when it was placed on the retail shelf.

Management was treating this issue in the same vein as if someone had recommended setting up a tee-shirt stand in the Sistine Chapel. The terminology used in the discussions is even symbolic of management's approach to the subject – the UPC label on the bottle's back panel was seen as a package 'violator' and was not to be considered, regardless of the cost necessary to apply it to the bottles' bottom panels.

So management had made two significant product packaging decisions that had substantial pricing implications to the product. The Lexan resin was probably a 50% price premium to PVC and the machinery to adhere UPC labels to bottle bottoms as opposed to their back panels would represent an upcharge of tens of thousands of dollars.

Who was missing from this decision-making process? Whose opinion was never sought at any time during this significant marketing interlude?

The customer for these products could have been asked some very basic questions that could have saved the organization in question literally hundreds of thousands of dollars. We can probably quickly formulate the questions we might want answered by our current customers, at least. 'Can you tell the difference between Bottle 'A' (Lexan) and Bottle 'B' (PVC)?' 'If yes, is the difference something that would impact your purchase decision?' 'If you can tell a difference between 'A' and 'B' and prefer 'A', how much more would you be willing to pay for 'A' versus 'B'?'

At this point the customers who might prefer one versus the other will tell you they would not pay more for the premium packaging. It is merely a 'nice to have', not a 'need to have'. Therefore, our organization needs to continually be in touch with our customers concerning their real needs and expectations for our products and services.

As a post-script to the Lexan plastic bottle example, several years later this plastic resin was replaced with a far less expensive resin for the manufacture of these bottles. Those recommending the change in resins were hailed for their significant 'cost improvement project', but was it really? It is true that bottle manufacturing costs were reduced, but wasn't it also just as true that they never should have gone up in the first place if we had really been listening to our customers?

We also want our customers to rank their needs and expectations for us. Clearly, when we are buying a product or service we have a hierarchy of expectations and needs. We now want to find out what these are for our products and services from our customers' point of view.

In the process of ranking these needs and expectations, we should also have our customers weight the relative importance of these ranked needs and expectations. For example, if our customer had ten needs and expectations that they had ranked from one to ten, how would they weight each item relative to each other? In other words, if they had to spread 100 points across these ten items, how would they do it? Would they give items one through ten each 10%? Or, for example, would the first two items be the most important and be weighted 30% and 20%, respectively?

These are all issues that professional customer satisfaction measurement experts can determine for us. In general terms, these are not rankings and weightings we want to do on our own. However, smaller organizations should not be adverse to discussing the relative ranking and 'general' weighting of needs and expectations with their customers, as we discussed earlier. It is not always necessary, or financially feasible, to have this measurement done by professionals. Again, qualitative discussion groups or simple surveys can certainly help start this process and provide even a very basic understanding of this important information. Importantly, larger organizations, with more demanding information needs and more intricate customer networks, should not believe that this preliminary information gathering can substitute long-term for professionally conducted customer satisfaction research.

It is important to understand that once we obtain this ranking and weighting data from whatever source, we have at our disposal very powerful knowledge about what is really important to our customers. It is also quantitative knowledge, not merely 'warm feelings'.

Among the World Class organizations in Customer Satisfaction, Federal Express is notable for its detailed knowledge of their customers' needs and expectations. This knowledge translates to a detailed ranking and weighting of needs and expectations, against which Federal Express can then measure customer perceptions of their performance on a *daily* basis.

Specifically, Federal Express knows the relative ranking and weighting by customer type of approximately 25 different needs and expectations. This is one of the Best Practices that we will discuss in detail in Chapter 5. However, it is important to realize that this ranking and weighting of needs and expectations is not theory. World Class organizations are practicing it and it is naive for our organization to try to compete in the twenty-first century without this type of customer information.

STEP TWO SUMMARY ACTIONS

Here are some summary actions to consider as we move to determine customer needs and expectations:

1. Determine our customers' needs and expectations from their point of view, not ours.
2. We can help them determine their needs and expectations more specifically by segmenting them by characteristics of products, services, and delivery;
3. Ask our customers to rank their needs and expectations from most important to least important;
4. Ask them to weight these ranked needs and expectations;
5. Calculate the five year value of each of our Top Ten Customers; determine what we are actively doing to keep them;
6. Develop an ongoing plan for increasing the number of customer contacts made by management personnel, following the Sam Walton example. Such contacts could include regular field visits, participation in sales calls, attendance at customer focus groups, customer forums, etc.

STEP THREE DETAILS – MEASURE CUSTOMER PERCEPTIONS

Armed with the knowledge of what customers need and expect from us, we can now measure their perceptions of our performance versus those expectations. We can also learn about their perceptions of our competitors.

We must always remember that the customer's perception of our performance is all that matters (Figure 4.4). We probably have all had the experience of perceiving a product or service in a certain light while the supplier saw things differently. What sense does it make for us to argue with our customers over the accuracy of their

Customers perceive service in their own unique idiosyncratic, emotional, erratic, irrational, end-of-the-day and totally human terms. Perception is all there is!

— Tom Peters

Figure 4.4 Customers perceptions are all that matters.

perceptions, when they are perfectly free to take their business elsewhere? There is no sense in winning an argument if we lose a customer. Our customer's perception must be our reality.

Yet, countless organizations have been trained to take just the opposite approach. It is common in the accounts receivables departments of many organizations to have department personnel interacting with their customers as if they were dealing with penitentiary convicts on work-release programs. Customers owing our organization money are perceived as dishonest thieves who are 'using our money', or 'float', for their own personal gain. Generations of accounts receivables clerks and managers have been trained to assume that the customer is purposely trying to delay payment for as long as possible. Customer inquiries concerning billings are often viewed merely as another tactic to delay or avoid payment entirely.

One Fortune 100 organization had followed this approach to accounts receivables until one of their senior financial managers decided to challenge the 'customer as thief' paradigm. Upon investigation with other senior managers in the department, it was determined that the vast majority of billing disputes were, in fact, the fault of the billing organization, not the recipient of the invoice! Detailed studies were conducted in which disputed invoices were carefully examined and tracked through the entire billing process, from start to finish. This study determined that in over 90% of the disputes, the billing organization was at fault.

This study emphasizes one of the basic Quality Improvement points mentioned in *Making Quality Happen* – the need to understand work as a process, to flow chart major organizational processes, and to examine problematic processes in step by step detail. The accounts receivables study demonstrated that the process involved was far more extensive than that solely under the purview of the account receivables department. Manufacturing, marketing, sales, and product distribution centers were all part of the process. For example, customers were billed for orders placed, but if the orders were not filled completely, then full payment was not made. Clearly, in our personal life, if we buy a dozen eggs we expect there to be 12 in the carton. However, if the store only has ten available today, then most of us will gladly pay for the ten now and the additional two when they are available. This is exactly how our customers feel, yet many organizations do not treat them that way.

Additionally, a whole host of other problematic opportunities surfaced during the examination of the invoicing-billing process, including the fact that invoices were often addressed to the wrong people, at the wrong addresses, with the wrong prices, or were just simply difficult to decipher. Again, these are not the customer's fault, they are ours. Frankly, this situation is common across all types of industries. In almost every organization examined, in a wide variety of fields, the majority of billing and payment problems trace their origin to the organization sending the bill, not the person receiving it and being asked to pay. And the real root cause of the problem is that the billing organization does not perceive the process from the customer's vantage point nor do they understand how the customer actually perceives it currently.

Therefore, we had better know exactly how our customers perceive our performance versus their ranked and weighted expectations.

Results from this type of study will clearly point out where we need to place our attention and where we may have been placing unnecessary attention (Figure 4.5).

We can also learn how our customers perceive the performance of our competitors. This can add an important element of urgency if we find that our competitors are outperforming us in the eyes of our customer on their priority needs and expectations. Additionally,

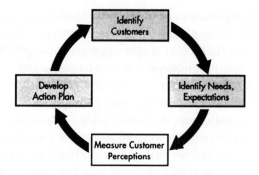

- Ask customers, 'How are we doing delivering what you want and expect for each characteristic?'
- Ask how they perceive the competition
- The shortfall between identified needs and perceptions provides the prescription for improvement.

Figure 4.5 Step three: measure customer perceptions.

this measure gives us a clear 'prescription' for where we need to take immediate corrective action.

MEASURE CUSTOMER PERCEPTIONS – THE PROFESSIONAL'S ROLE

Importantly, this type of measurement is ideally handled by Customer Satisfaction Measurement professionals. We are going to develop strategic plans based upon this information, so we want to make sure that it is accurate and actionable. This is clearly not an instance when we want to 'skimp' on details and quality.

The Customer Satisfaction Measurement professional's role will be discussed in detail in Chapter 6. However, it is important when we are discussing the topic of Measuring Customer Perceptions to understand the critical importance of 'third party' research.

It is quite possible for our organization to determine our customer's needs and expectations through direct interaction with that customer, as we have discussed previously. Clearly, if we are a large organization and need statistically projectable data, personal interviews are not the way to proceed. However, if we are a smaller organization, a non-profit group, or just beginning our journey to *Making Customer Satisfaction Happen*, it is appropriate to obtain this information in a qualitative manner, at least at the start of our journey to Customer Satisfaction.

We can also ask our customers to rank and weight their needs and expectations in this same information gathering process. We have to continually remind ourselves that this information is Qualitative, but it may well be suitable for our initial needs if we are aware of that *caveat*.

However, we run a serious risk of totally compromising our information if we then ask our customer, in the same face-to-face forum, to evaluate our performance versus these ranked and weighted needs and expectations. We are confronting an issue of human nature. Specifically, very few people have the personal confidence, temerity, and self-assurance to honestly evaluate our performance in a face-to-face situation. That is why most organizations have such terrible results in implementing personnel evaluation programs.

In the vast majority of cases, people are rarely honest with each other in face-to-face personnel evaluation situations. Therefore, we should not expect our customer to be completely honest with us. Nor should we be satisfied to receive incorrect, overly complimentary evaluations that might be the result of such interactions. These results will merely skew our information, misdirect our analysis, and seriously undermine our entire approach to World Class Customer Satisfaction. So while qualitative research is suitable for initial overviews of customer needs and expectations and their relative ranking and weighting, it is not appropriate for our organization to then ask customers to evaluate our performance versus these expectations. That type of critically important research is best left to quantitative research conducted by experienced professionals.

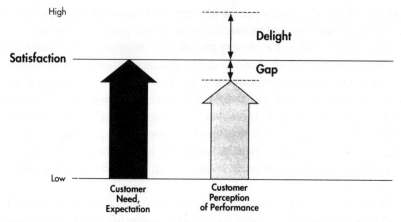

Figure 4.6 The perception gap: the difference between expectation and perception.

Quantitative measurement of the customer's perceptions of our performance will provide a clear picture of the 'Perception Gap' – the difference between our customer's perceptions of our performance and their needs and expectations (Figure 4.6). Determination of this gap has clear and direct strategic implications.

If we are perceived as performing below their level of expectations, then we need to address this area. If we are perceived as exceeding our customer's needs and expectations, we possibly have delighted some customers and we could consider marketing and merchandising our performance to potential customers and those currently with our competition. The entire subject of customer perceived 'gaps' with our product and service offerings is detailed in *Delivering Quality Service* by Zeithaml, Parasuraman and Berry.

IMPROVEMENT TARGETS FROM ACTUAL PERFORMANCE

Another important measure, which should be done concurrently with the measurement of customer's perceptions, is the measurement of our actual performance versus customers' prioritized and weighted needs and expectations. This measurement can be completed from our own internal sources and will provide us with some interesting insight on how our customers' perceptions match our actual internal performance.

For example, how long does it actually take for our customers to receive their order from the time *they* place it until *they* receive it? (Not from when *we* receive it until *we* ship it!) This is a classic argument concerning what constitutes 'next day' or two day delivery in some organizations that actually have 'guaranteed' delivery schedules. These organizations will state that they guarantee delivery in two days. However, the key question is when the two days start. Organizations usually believe that 'two day delivery' means that the customer receives the product or service two days after the organization mails, ships, or sends it (Figure 4.7).

Invariably, the customers' definition of 'two day delivery' is substantially different – they view the two days beginning at the moment they hang up the telephone after ordering the product or service. There can be a substantial difference between these two time sequences. We should know by this time that the only measure that matters in this operation is what the customer perceives, so our organization had best be measuring with the mileposts defined by the customer.

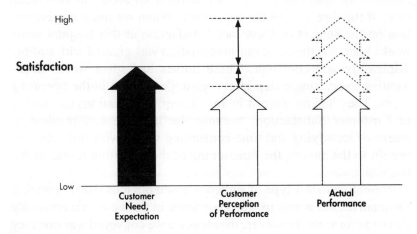

Figure 4.7 Setting improvement targets.

103

How long does it take us to invoice our customers? What is our invoice accuracy level? How many orders are sent out completely and correctly? How often do our sales representatives actually see our customers? This determination of our actual performance levels involves going into our organization and determining what our actual level of performance is. Clearly, the World Class organization already has this information because it forms the basis of the organizational measures that their management reviews monthly, if not weekly or daily.

However, the vast majority of organizations do not have this type of information readily available. Therefore, this step of determining our actual performance level for these key processes and services will provide us with invaluable insight and understanding concerning our organization. It is important to remember that we are not looking to develop performance levels for every operation in our organization. Rather, we are trying to determine actual performance levels on the processes identified by our customers as being important to them. There is no need to create work in pursuit of Customer Satisfaction, yet organizations can become confused.

For example, consider the case of the operating room nurses at a major hospital. They were advised by a consultant that Total Quality Management and Customer Satisfaction could only be achieved by a process mapping *every* process involved within their area of responsibility – the operating room. They dutifully set about this Herculean task, at the direction of the consultant. When we made a presentation on the subject of Customer Satisfaction at this hospital some weeks later, the theme of our presentation was greeted with audible displeasure and contempt. These nurses had spent weeks documenting every single step in every single process in the operating room. They had no interest in any further presentations on Quality or Customer Satisfaction, because the last one had resulted in weeks of stultifying and time-consuming work, with little obvious benefit to the nurses, the functioning of the operating room, or the hospital.

Presenting to this type of audience posed obvious risks: operating room nurses had access to, and knowledge in the use of, exceptionally sharp instruments. However, the message we conveyed was one they really needed to hear. The message was *focus*. Common sense should tell us to *focus* our efforts on Customer Satisfaction. Therefore, these nurses needed to focus on what was most important to their customers, including patients and the surgical staff. Every other process could wait. They needed to know their actual performance measures on those processes that were critical to their customers.

We cannot change customers' perceptions by addressing the perception directly. If people perceive American cars as lagging behind the Japanese in quality, then all the advertising to the contrary will not change that perception until we address the actual performance. We can only address the performance issue when we know where we are *today* – our baseline measure. From there, we can set improvement targets for specific and focused processes (Figure 4.8)

There is not one and only one measure of Customer Satisfaction. Rather, customers rate us on dozens of items and from their own list of priorities. Here we see the top four priorities (dark arrows) and what constitutes a level of satisfaction (dotted line) from the customer's perspective. Specifically, this line represents the performance level customers have said they want, need, and expect from us. The grey arrows show how customers perceive we perform on

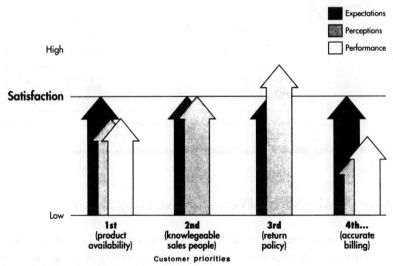

Figure 4.8 Planning for improvement.

each of the four identified needs and expectations. Obviously, we're doing fine on number 2, and even exceeding their expectations on number 3. But items number 1 and 4 are not reaching the customer's required level of satisfaction.

Those two areas that are trailing in customer perception should be our first areas of investigation – to determine just how we're actually performing. So again, we are focusing our efforts for improvement.

With these three important measures – what customers want, how they perceive our performance versus these 'wants', and how we're actually performing – we're ready to take improvement actions to close the satisfaction gap (Figure 4.9).

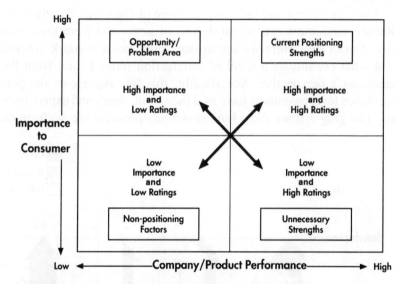

Figure 4.9 Gap analysis: opportunities exist in upper left quadrant.

GAP ANALYSIS DRIVES OUR ACTION PLAN

A potential output of a customer satisfaction study that will help prioritize our improvement actions could be this type of gap analysis matrix. On one axis (vertical) we measure the customer's view of the relative importance of various attributes of our products or services. On the other axis (horizontal) we chart the customer's perception of our performance on those attributes. Opportunities for improvement become very clear in the upper left quadrant where we are lagging behind customers' expectations.

Specific improvement opportunities exist in areas that our customers believe are very important but in which they perceive we are not performing well. This is even more enlightening if we develop a comparable matrix for our competitors and compare our customers' perceptions of us with their perceptions of our competitors. Do our competitors share the same profile, or are they poised to take our customers by being perceived more favorably in areas of greater importance?

Additional opportunities exist on the positive side, in areas that are both important to our customers and where they perceive we are doing very well in comparison to our competitors. These situations represent a clear marketing and promotional opportunity that we can leverage versus our competition in an effort to win over their customers.

The significant actions our organization takes on the basis of customer satisfaction measurement demonstrate the clear strategic

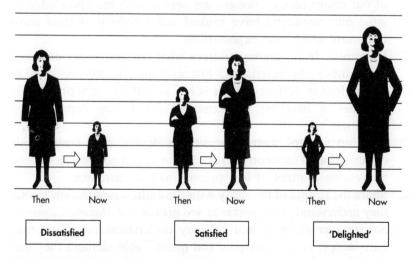

| Then | Now | Then | Now | Then | Now |

| Dissatisfied | | Satisfied | | 'Delighted' |

Figure 4.10 Measure customer perceptions.

107

importance of customer satisfaction. However, we will never know about these opportunities to improve or promote our customer-delighting performance if we do not have these quantitative results.

By measuring our customers' perceptions of our performance over time – and taking appropriate action where necessary – we can begin to move an increasing number of our customers into the 'delighted' category (Figure 4.10). We will no longer have to guess about what percentage of our customers fall into which category.

We will have quantitative data that demonstrates our customer's perceptions. And the more customers we move into the 'delighted' category, the more likely we are to retain them and build market share. Then we can begin looking at our competitors' customers and developing our strategic plans to make them our 'delighted' customers.

STEP THREE SUMMARY ACTIONS

Here are summary actions to consider in measuring customer perceptions of our products and services:

1. Determine how customers perceive our performance, and that of our competitors, once we are armed with the knowledge of what our customers have ranked and weighted as their most important needs and expectations.
2. Determine from internal sources what our actual performance is on these characteristics. For example, if we are seen as being slow in delivery, what is our *actual* delivery time? This provides a baseline measurement for future improvement activities.
3. Develop a 'gap' matrix, or similar device, that enables our organization to focus on the priority issues and not get sidetracked addressing matters of low priority to our customer.
4. Share the results of the study with the entire organization so that they understand the importance and primacy of *Making Customer Satisfaction Happen* and so they understand that Customer Satisfaction is a measurable and quantifiable element they can directly impact.

STEP FOUR DETAILS – DEVELOP ACTION PLAN

The final element of our *Making Customer Satisfaction Happen* model is to develop an action plan based upon the analysis we have conducted on the customer perception study results (Figure 4.11). Our gap matrix, or similar tool, should clearly identify where our efforts should be focused. We can then determine actual performance levels in these processes to serve as a baseline for future improvement efforts. Now a plan must be developed to turn these data into action and improve our performance levels until they exceed our customer's expectations.

In developing our action plan, it is critical to focus on addressing those needs and expectations gaps based upon their level of relative importance – ranking and weighting – to our customers. We do not

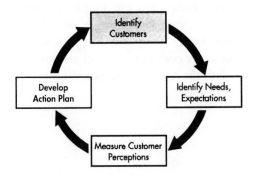

- Compare customer expectations with their perceptions (identify gaps).

- Based on customer priorities and gaps, select areas for improvement; measure actual performance levels for these critical areas.

- Improve performance levels until they meet or exceed customer expectations.

Figure 4.11 Step four: development action plan.

109

want to spend the organization's time and resources addressing areas which are not priorities to our customers, such as those in the bottom left and right quadrants of our gap matrix. Focus is vital.

Federal Express can immediately focus on specific problem areas as they occur. Federal Express management reviews their Customer Satisfaction results at their daily morning staff meeting. There is little or no lag time for small problems to fester into larger ones and for mildly dissatisfied customers to become angry customers telling 20 or more people about Federal Express' failings.

As we develop our action plan, we must certainly place initial emphasis on fixing the customer problems we know about through direct complaints, even if they represent potentially only 4% of dissatisfied customers (Figure 4.12). Moreover, we should also continually challenge the internal organization concerning our customer's perception of us. Specifically, are our policies and procedures focused on satisfying our customers and on assisting those who touch our customers to satisfy them? Do our employees know, as do Fred Smith's at Federal Express and those of the Ritz-Carlton Hotels, that they are empowered to do anything necessary to satisfy the customer? Our customer contact personnel are our first line of defense – and offense – with the 4% of dissatisfied customers who actually

} 4%

For the 4% who complain, customer complaint personnel who:
- Are accessible
- Listen and ask questions
- Empathize and understand frustration
- Are good problem solvers
- Are empowered to act on the customer's behalf

Figure 4.12 1. Fix customer problems.

contact us. We need to make certain they are fully empowered to satisfy our customers.

Do we stand behind our products and services with our customers? Do we have a customer-oriented policy or approach to the traditional Customer Service function? What kind of guarantees do we offer our customers?

L.L. Bean, the family-run mail order outdoor clothing company with sales of approximately $600 million, has a Customer Satisfaction guarantee that is unequivocal and unsurpassed:

> Our products are guaranteed to give 100% satisfaction in every way. Return anything purchased from us at any time if it proves otherwise. We will replace it, refund your purchase or credit your credit card, as you wish. We do not want you to have anything from L.L. Bean that is not completely satisfactory.

That is a company policy focused on the customer. Why couldn't our organization merely substitute our name for that of L.L. Bean? Does it really matter what business or occupation our organization is involved in? This is a model for any type of organizational policy statement. Ours should be just as clear, concise, and comprehensive (Figure 4.13).

Dissatisfied   Satisfied

- Share customer data with all employees
- Develop a 'customer focus' among all employees
- Set standards for key processes based on customer needs and expectations
- Implement improvement teams to bring key processes to standard
- Measure, measure, measure

Figure 4.13 2. Prevent dissatisfaction.

While each action plan will differ by organization, there are some basics for any action plan that bear mentioning.

We want to move those customers who are currently dissatisfied to the ranks of the satisfied.

We start this by focusing the organization on World Class Customer Satisfaction. One way to do that is to share our Customer Satisfaction measures with the entire organization. Remember, what gets measured, gets done.

We should set both short and longer term improvement objectives in the areas we have identified for action. We should not have to wait one year to see if we are really improving. This does not mean that we need to do a major customer satisfaction study each month or have the daily data input of Federal Express.

It does mean that we can measure our performance in those areas identified by our customers and see if we can demonstrate significant improvement. For example, if our customers say our delivery time is too long and we know from our internal data the actual delivery performance, are we seeing any improvement and shortening of that delivery cycle based upon our action plan? Additionally, we can have informal, interim measurement from our customers by asking them if they have seen improvement.

We should make sure our standards are realistic and consistent with our customers' needs and expectations. We do not win any battles for satisfied customers by overpromising and underdelivering. This is a classic instance where our personal life experiences should guide our professional life practices. How often have we read or heard about some product or service that makes great promises and then fails to deliver? What about those product or service guarantees that sound so enticing, but when called upon to produce, fall far short of expectations? A good friend once summarized the business destroying potential of overpromising by stating, 'Beware of individuals who promise the World. They usually end up delivering Columbus, Ohio!' Nothing wrong with Columbus, Ohio, but it is about six continents less than what we were promised!

We need to adjust our internal standards to those of the customer and continuously measure our performance in this light. Our old internal standards – 'we've always done it this way' – are no longer relevant and should be discarded. From its inception, our strategic focus on Customer Satisfaction makes the customer our arbiter, not our internal organization.

We should guarantee that our improvement plan is clearly defined and that the organizational elements involved clearly understand the reason for the necessary improvement and their role in making it happen.

As we stated earlier 'satisfied' customers are not enough. We want all our customers to be 'delighted'. Our entire organization needs to know that this is our number one priority (Figure 4.14).

This priority on customer 'delight' relates directly to the need to communicate our action plan goals to the entire organization and to inform them about our progress towards achieving these goals. This widespread communication reinforces the importance of customer delight and builds employee enthusiasm and participation by clearly showing them measurements of improvement against our stated goals.

We also need to continually guarantee that our organization is structured to really address the needs of our customer and the needs of those who directly deal with them. We will discuss this important subject at length in reviewing Best Practices in Customer Satisfaction in Chapter 5, but here are some basic elements to address.

Satisfied   Delighted

• Make 'delighting the customer' your organization's top priority
• Examine existing policies and procedures for their customer 'friendliness'
• Free employees to delight customers
• Reward and recognize employees for instances of outstanding service
• Measure some more

Figure 4.14 3. Beyond satisfaction.

113

We certainly need to examine our internal policies for their customer 'friendliness'. Do we have a stated (or unstated but clearly practiced) 'no return' policy? One division in a Fortune 100 company had such a policy and prided itself on the fact that they never took any product back! How does that approach contrast with that of L.L. Bean? Which organization would we rather deal with if we had a choice? Which organization do you believe is really customer-focused and which organization do you think has customers 'singing' its praises to other potential customers?

Are our employees free to really satisfy the customer or are their hands tied by 'red tape' and internally focused policy and procedures? Nordstroms wants to know why employees said 'no' to customers, rather than why they said 'yes'. L.L. Bean will take any merchandise back, with no questions asked. At Federal Express, couriers are empowered to spend as much as $250 to satisfy the customer (the average Federal Express package costs $25), while Customer Contact personnel are empowered to credit accounts up to several thousand dollars without higher level approvals.

As Tom Peters illustrated in his video, *A Passion for Customers*, at least one Federal Express employee even went so far as to rent a helicopter to make certain that phone lines were open during a severe ice storm in the far western section of the United States. That personifies empowerment and exemplifies employees who are focused on delighting customers our organization recognizes as having a long term value to our success.

These procedures also speak of organizations that value and cherish their employees and realize that employees are their first line of contact with the customer. Their customer focused and employee enhancing procedures call to mind a quote from former United States Secretary of State and advisor to many presidents, Henry L. Stimson. Stimson stated that the best way to make someone 'trustworthy' was to trust them. These World Class Customer Satisfaction organizations have heeded those sage words and applied them directly to their employees.

A key part of our action plan is Process Improvement (Figure 4.15). We discussed in Chapter 3 that our organization 'touches' our customer during thousands of 'Moments of Truth' each day. Some process is involved at each of these 'Moments of Truth'.

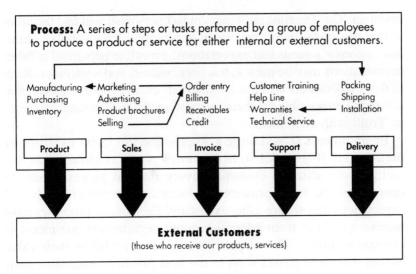

Process: A series of steps or tasks performed by a group of employees to produce a product or service for either internal or external customers.

Manufacturing	Marketing	Order entry	Customer Training	Packing
Purchasing	Advertising	Billing	Help Line	Shipping
Inventory	Product brochures	Receivables	Warranties	Installation
	Selling	Credit	Technical Service	

| Product | Sales | Invoice | Support | Delivery |

External Customers
(those who receive our products, services)

Figure 4.15 Process improvement.

For example, if an airline attendant serves us an unacceptable meal on our next flight, all the apologies, free drinks, and 'niceness' will probably not erase the perception made by that bad meal. Further, the flight attendants themselves can really do nothing about that bad meal, just as they can do nothing about flight delays, late arrivals, or lost luggage. That does not mean they will not be blamed for each of these because they are on the 'frontlines' with customers.

The responsibility – the 'ownership' – of the in-flight meal resides somewhere else in the organization, such as the kitchen or the meal planner's desk. The meal preparation process – or possibly meal delivery and/or storage process – needs to be addressed and improved. This issue will need to be addressed by employees who probably never see the customer, but they have a direct impact upon them, as each of us clearly knows.

So we see that we have a 'chain' of internal customers and suppliers who are critical to the organization's goal of delighting our external customer. The sales person taking the order may be the one in direct contact with our customer, but those in manufacturing making the product, those in packing and shipping sending it to our customer, and those in billing, all directly influence that customer. Our frontline customer contact personnel will feel the brunt of that dissatisfaction if anyone in our internal 'chain' fails to meet our customer's needs and expectations. Frontline personnel in other organizations may be nurses, teachers, waiters and waitresses. Each of these individuals also has an 'internal' chain supporting them that is directly responsible for their 'success' or 'failure' in their Moments of Truth with customers.

Every process should forge a 'satisfaction link' with its customer organization (Figure 4.16). In delivering products and services that 'delight' our external customer, every internal process needs to operate on the same principle of exceeding customer needs and expectations that we are using externally. Each of our processes must receive an output from internal supplier departments that meets or exceeds its needs. That department, in turn, performs their value adding work and passes it on to the next function, exceeding their needs and expectations.

The external customer will be the ultimate beneficiary of this commitment to *Making Customer Satisfaction Happen*.

The entire subject of Process Improvement is extremely important and is often referred to as 'reengineering'. It is not our aim to cover that subject here. The entire subject of reeingineering is

Process Improvement Steps:

- Document (map) the steps of the process

- Develop standards for the output based on customer needs

- Measure the output against the standard

- Make process improvement until the output meets standards

Figure 4.16 Forging the links.

covered in depth in *Reengineering the Corporation* by Michael Hammer and James Champy. Their approach focuses on differentiating reengineering from basic improvement. Reengineering really involves major restructuring or process innovation. Basically, it embraces our 'common sense' approach and asks, 'if we had to design this process from scratch, what would we do?' However, we can summarize some of the key elements of this important approach, particularly pertaining to our action plan for Customer Satisfaction. A more detailed discussion of this 'process' approach is detailed in Chapter 6 of *Making Quality Happen* where we discuss its importance under the topic 'If we learn only one thing about Quality'.

We can form Process Improvement Teams in forging the links of our internal processes. These teams comprise those individuals directly involved in the process under review. Importantly, we must remember to focus our action plan efforts on processes prioritized by our customers' ranked and weighted needs and expectations, so keep the number of teams to a minimum at the outset. These teams may want to consider initial process improvement actions before jumping ahead to total reengineering, which may require extensive organizational resources and time, but is clearly worth the effort in the long-term. These teams should consider the following initial steps, at a minimum:

1. Document (map) the process – where does it start, where does it end, and what steps occur in between;
2. Develop output standards based on customer needs – what does the customer want the output to look like and therefore how do these needs and expectations get translated into standards for our organization;
3. Measure process output against our developed standards – what do the actual outputs of our process look like, i.e. how long does it take our packages to arrive, how long does it take to replace a hip, launch an aircraft? etc.;
4. Improve process until the product or service meets standards.

BENCHMARKING'S ROLE IN ACTION PLANS

Once we have identified the internal processes that we need to improve to exceed our customers' needs and expectations, we

117

must learn more about these processes. Of course, we gained an important piece of process learning in just determining (measuring) the actual performance of these processes.

Benchmarking can provide another very enlightening piece of information about these processes. Benchmarking is comparing our processes and practices with the 'best in class' – the best practitioners of this process, regardless of industry (Figure 4.17).

For example, a hospital determined that food service was a very important issue with patients. Moreover, patients evaluated the hospital's food service very poorly. In an effort to improve this process, the hospital benchmarked the food service process with the 'best in class' practitioner – the Baldridge Award winning Ritz-Carlton Hotel group, which also serves large numbers of meals on a daily basis. The key to examining 'best in class' practices is that it elevates the comparison above our normal vantage point – our own industry or community of knowledge – and forces us to look at a process in a potentially new and different light.

Benchmarking is a major topic itself and is clearly and concisely discussed in a variety of Xerox Corporation *Leadership Through Quality* brochures and in Robert Camp's informative book *Benchmarking*. However, a few benchmarking points at this juncture are relevant to our *Making Customer Satisfaction Happen* discussion.

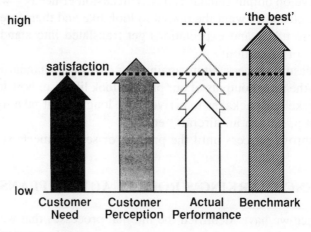

Figure 4.17 Benchmarking.

We are most effective in benchmarking if we narrow the focus of the process we are seeking to improve. For example, benchmarking a sales force would not be very helpful. What part of the sales process is it that we now need to address based upon the results of our customer satisfaction measurement? What aspects of a sales force has the customer identified as most important to them? Is it their product knowledge, their ability to handle billing/financial issues, their use of samples, or their call frequency? Once we have narrowed the focus of our process inquiry, we can then learn who does this best, both in our industry and who is the overall 'best in class'.

In developing *Making Customer Satisfaction Happen*, several specific Customer Satisfaction processes were benchmarked with 'best in class' practitioners. Their practices are presented in Chapter 5 and should serve as a *starting* point as we begin to consider their application to our organization.

Most organizations, particularly those that are 'best in class', are very willing to discuss their processes with us, so long as we are willing to share some 'best practice' from our organization. By learning from the 'best' we can shorten our process improvement cycle time and thereby move quickly to start exceeding our customers' expectations. That is the benefit of benchmarking – it is a tool to help us achieve our goal of improved Customer Satisfaction.

Once our organization has this very clear understanding of our customers' needs and expectations concerning our products and services, we can begin to use this information to not only improve the processes impacting our customers, but we can also begin to improve the existing and new products or services we are offering our customers.

The process for utilizing this customer information in the improvement of existing products and the creation of improved new products is called Quality Function Deployment. Again, this topic is an entire subject itself detailed in J.R. Hauser and D.P. Clausing's seminal article in the May–June, 1988 *Harvard Business Review*, but a few points about Quality Function Deployment are relevant.

Quality Function Deployment is a cross-functional system for translating customer needs, expectations, and requirements into organizational requirements. The output of Quality Function Deployment is better new products and services, developed faster and with fewer errors because the 'Voice of the Customer' has been integrated and built into the design.

Quality Function Deployment has been extensively used in the automotive industry worldwide. Here the prioritized needs and expectations of the customer have been translated by cross-functional work groups – Cadillac calls this 'simultaneous engineering' – into a finished product designed to 'delight' the customer.

A classic example of this type of effort is the Lexus from Toyota. This luxury car was designed to compete with the Mercedes, BMW, and Jaguar at two-thirds their price. The car was developed in two years – GM's Saturn took six – and Lexus became the most highly rated automobile in customer satisfaction its first year on the market – exceeding ratings for the cars it was designed to compete against.

The Lexus story is not an overly complicated one, but it is clearly an example of an organization that focused like a laser beam on its target customer and translated their needs and expectations, through tools like Quality Function Deployment, into a marketplace success, and did it in record time.

The critical part of the Lexus story was their willingness to listen to their customer – to 'reengineer' the process of developing a new car rather than merely improving upon the existing process. Importantly for those competing against Lexus, Toyota had no 'secret' weapon, no 'inside' information, no industrial espionage or wonder weapons. They merely brought an open mind and attentive ears to the voice of their potential customers. The result was unprecedented levels of customer satisfaction, marketshare gains, and revenue growth. Our organization can achieve these same results if we bring the same elements to bear in our action plan that Lexus did – an open mind and attentive ears to the voice of the customer.

STEP FOUR SUMMARY ACTIONS

Here are some summary actions to consider as we address the development of our Customer Satisfaction Action Plan:

1. Use the quantitative results of our Customer Satisfaction Measurement to prioritize our actions.
2. Address the needs and expectations to be improved based upon their customer-perceived importance.
3. Examine the business processes involved in the areas to be improved – look for 'root causes' (poor in-flight meals are probably not the fault of the flight attendant).
4. Gain employee input on process improvement opportunities – 'ask the experts' and consider reengineering problematic processes rather than merely improving them.
5. Provide the organization with clear, prioritized improvement plan objectives and also provide the required resources to accomplish these objectives.
6. Measure results on an ongoing basis. Do not wait until the next major customer satisfaction measurement.

SUMMARY LESSONS LEARNED

1. Our *'Making Customer Satisfaction Happen'* four-part model encapsulates our approach to integrating Customer Satisfaction into our organization.
2. We need to determine who our customers currently are, who they could potentially be, and who our competitor's customers are.
3. We need to prioritize our customers based upon criteria agreed upon by our entire organization – for example, actual revenue generated, potential revenues, industry impact, leverage, etc.
4. We need to determine our prioritized customers' needs and expectations and be able to rank and weight these items.
5. Our organization needs to understand and know how our customers perceive our performance and that of our competitors on addressing–delivering these needs and expectations.
6. Our organization must integrate into our strategic plan an action plan aimed at addressing the difference – the 'gap' – between our organization's actual performance level and those required by our customers.

REFERENCES

[1]   Peters, Thomas, J. and Waterman, Robert H. Jr. (1982) *In Search of Excellence*, Harper & Row, New York, p. 122 and Peters, Thomas J. and Austin, Nancy (1985) *A Passion for Excellence*, Random House, New York, pp. 8–34, 378–413.

[2]   Huey, John, 'How Sam Walton Does It', *Fortune*, 23 September 1991, pp. 46–59.

5

Best practices in *Making Customer Satisfaction Happen*

The purpose of business is to get and keep customers
Theodore Levitt

The *Making Customer Satisfaction Happen* Model presented in Chapter 4 encapsulates our approach to Customer Satisfaction. But have any organizations really incorporated this approach into their standard operating procedures? For instance, have organizations really differentiated their various customers and actually ranked and weighted the needs and expectations of each customer group? Has any organization ever actually surveyed their customers and obtained quantitative feedback on the customers' perceptions of the organization's performance? And have any organizations really developed workable action plans based upon these customer studies?

Our fondest wish is that everyone reading this list of questions could answer affirmatively for their organization. However, experience shows that approximately 90% will not have completed even one full circuit of the Model and probably over 99% have not structured their organization for maximum customer contact and impact nor have they made customer satisfaction the focus of their reward and recognition process.

There is 'good news' and 'bad news' in the content of this chapter if we find ourselves among the majority of organizations who haven't started on an organized and systematic approach to our Customer Satisfaction journey. The 'good news' is that there are organizations that have incorporated all aspects of our Model into their strategic focus and we can learn from them without having to suffer the trials and tribulations of working our way up the Customer Satisfaction 'learning curve'.

The 'bad news' is that there are organizations who have already made Customer Satisfaction their strategic focus and they have an exceptional head start on us. We can only hope that they are not our direct competitors and that we can get started immediately and thereby have an advantage over our competitors to the same degree that these 'best in class' practitioners currently hold over us.

If *Making Customer Satisfaction Happen* is to be a reality in our organization, the responsibility will rest with management. In every example we will examine, management leads the way and focuses the organization on customer satisfaction. Federal Express, Toyota, L.L. Bean, Nordstrom's, SAS, Milliken, all have managements obsessed with satisfying the customer.

Each of these organizations, and all the other World Class Customer Satisfaction organizations, know their customers' needs and expectations in incredible depth and breadth. They use their knowledge to improve their organizations, their products and their services to make them even more capable of delighting the customer.

And it is a continuing process. It is a continuous improvement process because World Class Customer Satisfaction organizations know that the standard is always changing. The 'bar' is always being raised. There are new competitors waiting to take their places at the top and waiting – even planning – to take their customers. So these Customer Satisfaction benchmarks are themselves not content with merely satisfying their customers. Their goal is to delight their customer at every opportunity – the Disney organization calls it the 'Wow' factor. Our objective should be nothing less.

A Customer Satisfaction benchmark organization has been selected for each of the major practices discussed throughout this book. These are certainly not the only organizations that could serve as benchmarks, but they are among the best known and have been the most consistent over time. Their Best Practices in specific areas of Customer Satisfaction demonstrate that the various elements we have discussed are not only achievable, but are being accomplished, and improved upon, as we read this.

124

The Best Practices to be reviewed are:

Strategic Focus for Customer Satisfaction
Organization Structure for Customer Satisfaction
Customer Expectation Identification
Customer Satisfaction Measurement System
Frontline Personnel Empowerment
Customer Satisfaction Guarantee

STRATEGIC FOCUS FOR CUSTOMER SATISFACTION

Federal Express

The Federal Express corporate philosophy is **people–service–profit**. Federal Express (FEDEX) chairman Smith states, 'When people are placed first, they will provide the highest possible service, and profits will follow'. The annual FEDEX corporate performance goals, upon which all bonuses are based, derive from quantifiable objectives in each of these three areas. Failure to attain these stated objectives in *any* of these three areas means no one in the organization, from the chairman down, will receive a bonus. In fact, the year after Federal Express won the Malcolm Baldridge National Quality Award they did not hit one of their targets and bonuses reflected that miss.

The **people** segment is measured through the FEDEX leadership index component of the company's annual employee attitude survey, a copy of which is included in the Appendix following Chapter 6. This leadership index is a statistical measurement representing subordinates' opinions of management's leadership performance. Chairman Smith has stated, 'Customer satisfaction begins with employee satisfaction'.

The **service** goal is measured through the *daily* average statistical measure for service quality and Customer Satisfaction in 12 critical Customer Satisfaction categories – determined by the customer – and totaled as the FEDEX service quality indicators. Most recently, FEDEX achieved a 94% 'completely satisfied' (top box) rating from its customers.

125

The **profit** goal is a percentage of pretax margin, determined by the previous year's financial results.

The corporate strategic commitment to Customer Satisfaction is reinforced in the two overriding business objectives at FEDEX: 1. 'We are committed to delivering each shipment entrusted to us on schedule 100% of the time. Equally important, we will maintain 100% accuracy of all information pertaining to each item we carry'; 2. Our objective is to have a 100% satisfied customer at the end of each transaction'.

Could our organization credibly make this type of commitment, regardless of industry? If we could not, it is time to begin planning on how we could and to begin this planning process quickly! [1]

Toyota–Lexus

Toyota Motor Sales, USA and its Lexus Division operate based upon separate but shared Customer Satisfaction focused strategic approaches. The deployment of these approaches differs given the difference in size between the two organizations (Toyota – one million annual auto sales, Lexus – 94 201 auto sales in 1993).

The annual sales objectives for both organizations are to be achieved through a strategic focus on Customer Satisfaction. As we discussed in Chapter 2, this Customer Satisfaction focus takes three strategic forms at Toyota–Lexus: 1. The retention of satisfied Toyota and Lexus customers. Currently 93% of Toyota owners with positive sales and service experiences will strongly consider repurchasing a Toyota and from the same dealer; 2. Positive 'word of mouth' endorsements from satisfied customers to potential customers – satisfied customers are seen as enormously impactful emissaries to potential customers by Toyota; 3. Additional new customer volume generated by Toyota marketing and sales efforts focused on customer satisfaction and product quality.

This strategic approach is explained diagrammatically in Figure 5.1.

The Business International corporation discussed the Toyota approach at Lexus in their research report on Maximizing Customer Satisfaction. 'When Toyota Motor Corporation first set its sight on

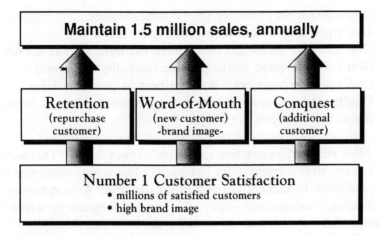

Figure 5.1 Strategic approach: Toyota.

the luxury car market, it knew it had to do more than just build another quality car. It had to give full attention to the ownership experience. "Simply put", says Richard Chitty, corporate manager for Service, Parts and Customer Satisfaction at Lexus headquarters in Torrance, California, "the world has long bought Toyota for its quality, but now we knew we had to give our customer something more".'

'Lexus, Toyota's entry into the higher-end car market, had an impressive debut. In the first six months after its introduction in 1989, the car nearly outsold BMW and resoundingly beat Porsche, Jaguar, Audi, and Sterling combined. How do Toyota executives account for their success in the competitive luxury car market? A strategic focus on Customer Satisfaction and Employee Empowerment is a major reason'.

'"For the first one hundred dealerships opened in the US, we picked the best of the best", says Chitty. (To select the successful 100 out of a field of 1 500, Toyota scrutinized customer ratings of dealer performance and evaluated dealers' track records in their chosen regions.) These dealers then came together and worked as a team to set standards for a new level of customer-service satisfaction. The following were some of their guidelines and goals':

- Don't just meet customer needs; exceed them.
- Don't just answer questions; anticipate them.
- Don't just be thorough; check even the most minute details.
- Don't just do good work; provide unparalleled service.
- Don't just do your best; do your best as part of a team.
- Don't just be courteous; make every customer feel as if he or she is the only one.

'"We've taken a proactive approach", says Chitty. "This means that every staff member who has any contact with a customer has the authority to make a correction on the spot. A receptionist, a technician, a salesperson – they all know that they can do whatever it might be they need to do to make the customer happy".'

'"This freedom, in turn, has produced not only happy customers, but also some very happy Lexus dealers", says Chitty. "Dealers are telling us this is the greatest time they've ever had in the car business because of the control and responsibility they feel. No one is limited by corporate bureaucracy; they can do what they think is best to satisfy their customers".'

'Employee empowerment certainly helped the newly launched Lexus save face in a time of trouble. In December 1989, Toyota had to recall the Lexus to fix a cruise-control defect – once turned on, the control could not be turned off. Although the problem was reported by only one driver, Toyota had to check out every single Lexus because such a defect posed a potentially serious hazard'.

'The recall could easily have turned into a marketing nightmare for the recently introduced luxury sedan, which cost upward of $38 000. The Toyota solution? Each owner received a personal phone call from the dealer, who offered the owner two choices: bring in the car yourself to be checked out and fixed on the spot, or have Lexus make a house call to pick up the car in the evening and return it ready to go the next morning. Most dealerships made their own decisions about whether to go above and beyond the recall repair: wash the car, fill it with gas, clean the interior, etc. Some dealers even placed a small gift, such as a box of chocolates or an ice scraper on the front seat'.

'Owners were mollified and more than a few were impressed. "They went the extra mile, more than any other manufacturer, to make sure I was completely taken care of and not inconvenienced", says William Beatty, a St. Louis doctor whose family owns three Lexus cars. "Nobody before had ever filled my tank or washed my car when it was taken in for repairs".'

'Comments Chitty: "All eyes were on us. It was Toyota's first luxury car, and it was introduced in the US first because the feeling was that the US was the best place to do the launch. All the dealers understood what this meant. In the end, I think what could have been a very bad situation for us ended up working to our advantage. It showed the public exactly what our commitment to Customer Satisfation is all about".' [2]

Source for material appearing on pages 126–9: *Maximizing Customer Satisfaction*, December 1990, Toyota case study, pp. 41–42, the Economist Intelligence Unit. [3]

ORGANIZATION STRUCTURE FOR CUSTOMER SATISFACTION

Toyota-Lexus

The Toyota Customer Satisfaction committee is Toyota's primary tool for accomplishing its goal and strategic focus of being the leading automobile company in Customer Satisfaction. The committee is organized to channel customer satisfaction information accurately and concisely to Toyota management so that Customer Satisfaction issues are handled expeditiously and effectively to benefit Toyota customers.

The Customer Satisfaction committee comprises four subcommittees representing every major operating department at Toyota: sales and marketing, parts and service, product quality, and Lexus. Through this process ultimate responsibility for Customer Satisfaction is assigned to all operational areas. This subcommittee structure breaks down traditional barriers and allows open communications to initiate effective customer satisfaction action plans. To accomplish this task, each subcommittee has developed long-range plans which will contribute to the whole. These individual plans are outlined below:

Sales and marketing – Achieve industry leading sales satisfaction, which will maximize initial customer satisfaction as well as capture additional repeat and conquest sales.

Parts and service – Achieve industry leadership in dealer after-sale customer handling and to establish a national reputation for quality and integrity in vehicle service.

Product quality – Achieve both industry leading product quality and product acceptance, which will maximize initial customer satisfaction as well as capture additional repeat and conquest sales.

Lexus – Achieve unequaled customer satisfaction by placing the customer first and making certain that each Lexus individual's day-to-day performance is greater than the customer's expectations.

Each subcommittee is chaired by Toyota vice-president level executives. These people are responsible for involving all affected departments in determining problems and developing action plans related to Customer Satisfaction. Involvement by upper management secures support and commitment for programs that address customer satisfaction issues.

The 'Voice of the Customer' is reported regularly to the subcommittees from the Customer Relations Department. This 'voice'

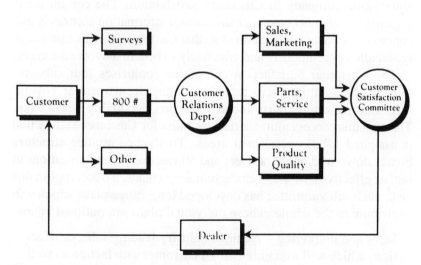

Figure 5.2 Best practices: organizational structure – Toyota.

130

takes the form of information from the three ongoing Toyota Customer Satisfaction studies (new vehicle sales, service, and overall satisfaction index) as well as 800 number (free call) input.

The subcommittees play a major role in gathering, analyzing, reporting, and distributing actionable information that can be passed on to the appropriate departments.

The Customer Satisfaction committee meets monthly, with each sub-committee reporting on a quarterly basis. These regular meetings provide management with frank assessments of customer satisfaction issues and actions undertaken by the departments within the subcommittees.

The Customer Satisfaction committee process is outlined in Figure 5.2.

CUSTOMER EXPECTATION IDENTIFICATION

Federal Express

Federal Express has been able to clearly identify, rank, and weight customer expectations for each of the wide variety of customer segments which it serves.

The FEDEX extensive Customer Satisfaction survey program – five separate surveys of different segments annually – guarantees that these customer expectations are continually updated regarding not only their relative rank and weight in importance to customers, but also concerning customer's perceptions of FEDEX's performance against these attributes and the performance of its competitors.

The Customer Satisfaction study includes 2400 telephone inter-views quarterly comprising: 900 US Domestic Base Business, 900 US Export, 300 Drop Box, and 300 Service Center. Respondents are asked their satisfaction with FEDEX and its competitors on 53 total attributes subdivided to 17 attributes for US Domestic customers, 22 attributes for US Export customers, six attributes for Drop Box customers, and eight attributes for Service Center customers. While it is unlikely that the relative ranking and weighting of specific attributes will change dramatically over time, FEDEX is clearly pro-active in this area given the continual nature of this survey with its daily calls to customers.

131

The various attributes for each part of the Customer Satisfaction Study are outlined below and on the following pages, as is an example of the customer attribute ranking and weighting developed by FEDEX.

FEDERAL EXPRESS CUSTOMER SATISFACTION ATTRIBUTES MEASURED
- • Total Attributes Measured
- • 17 US Domestic Attributes Measured

- Overall satisfaction
- Ease of doing business
- Value for the price
- Billing
- On time delivery
- Condition on arrival
- Tracing
- Phone service agent
- On time pick up

- Ordering-receipt of supplies
- Pricing/rates
- Geographical coverage
- Phone answer promptness
- Saturday service
- Couriers
- FEDEX account executive
- Airbill

Customer satisfaction study 22 US export attributes

- Overall satisfaction
- Customs clearance
- Handling of duties and taxes
- Service to Canada
- Service to Puerto Rico
- Service to Far East
- Service to Great Britain
- Service to European continents
- Pricing/rates
- Ease of completing international paperwork
- Customer service
- On time delivery
- Speed of delivery
- Billing
- Tracing
- Account executive
- Courier
- Geographic coverage
- Value for the price
- On time pick up
- Condition of package on arrival
- Proof of delivery

Customer satisfaction study

SIX DROP BOX ATTRIBUTES

- Overall satisfaction
- Proximity of drop box to home or office
- Accessibility/visibility of drop box
- Availability of supplies and packaging materials
- Information specifying locations of drop box
- Appearance and condition of drop box

EIGHT SERVICE CENTER ATTRIBUTES

- Overall satisfaction
- Proximity of center to home or work
- Operating hours
- Ability of customer service agent to explain service features and benefits
- Helpfulness and courtesy of customer service agent at center
- Accessiblity/visibility of center
- Availability of supplies and packaging materials
- Speed of service

Importance ratings

MAY

Rank	Attribute	% Rating 95–100
1.	On time delivery	91.6
2.	Condition of package on arrival	86.9
3.	On time pick up	81.3
4.	Tracing	78.1
5.	Ease of doing business	76.4
6.	Value for the price	76.0
7.	Handling of damaged/ lost shipment claims	75.3
8.	Geographic coverage	71.9
9.	Agents handling trace	70.2
10.	Providing information	69.0
11.	Customer service agent on the phone	68.7
12.	Personnel who handle billing	64.5
13.	Pricing	63.0
14.	Federal Express answering phone promptly	62.3
15.	Couriers	60.2
16.	Customer service agent at Federal Express facility	59.3
17.	Preparation of packages for shipment	57.2

High importance: ranks 1–7

Moderate importance: ranks 8–17

	Rank	Attribute	% Rating 95–100
Moderate importance	18.	Billing	55.7
	19.	Airbill	54.0
	20.	Receipt of supplies	52.8
	21.	Credit/collection procedures	52.2
	22.	Federal Express account executive	51.1
Low importance	23.	Ordering supplies	47.1
	24.	Service guide	43.9
	25.	Declared value	40.2
	26.	Saturday service	36.6
	27.	Constant surveillance service	29.7
	28.	Restricted articles/hazardous materials	25.7

CUSTOMER SATISFACTION MEASUREMENT SYSTEM

Federal Express

A commitment to clear, frequently repeated, and universally understood Customer Satisfaction goals – followed by continuous measurement of progress against these goals – forms the foundation of the Federal Express approach to the Measurement of Customer Satisfaction.

The company's two ambitious Customer Satisfaction goals are straightforward and unequivocal: 100% Customer Satisfaction after every interaction and transaction, and 100% service performance on every package handled.

Importantly, Federal Express has developed a Service Quality Indicator (SQI) which is a 12-item statistical measure of Customer Satisfaction and Service Quality from the *customer's* viewpoint. Federal Express tracks these 12 items, individually and in total, *every day* and they are reviewed by the company's top management at their daily meeting, first thing each morning. They are reported weekly to all FEDEX employees worldwide via the company's in-house television system FXTV. These 12 items, and their relative importance to customers, are outlined on page 139:

138

Lost Packages	10 points
Damaged Packages	10 points
Missed Pick ups	10 points
Wrong Day Late Delivery	5 points
Lost and Found	5 points
Complaints Reopened	5 points
Abandoned Calls	1 point
Invoice Adjustments Request	1 point
Traces	1 point
Right Day Late Delivery	1 point
Missing Proof of Delivery	1 point
International	1 point

Each year the company's service goal (people, service, profit) has been set based upon a degree of progress towards the company's five-year goal of a 90% reduction in total SQI levels from its initiation in 1988, regardless of increased package volume.

In addition to this daily measure of Customer Satisfaction, FEDEX spends over one million dollars each year on five major Customer Satisfaction surveys of its various customers. These studies are outlined briefly below:

1. Customer Satisfaction Study – This is a quarterly telephone survey of 2400 customers, chosen at random, with phone calls conducted on a daily basis. This survey covers four market segments: 1. base business – the average FEDEX customer; 2. US export customers; 3. manned-center customers (those who drop packages at FEDEX owned store-front service centers); and 4. drop box customers.

   During a ten-minute interview each customer's satisfaction is gauged on a five-point scale against a 53 item list of service attributes. FEDEX has chosen to use only 'completely satisfied' (top box) responses as its measure of Customer Satisfaction. (In a recent study, the company achieved a 94% 'completely satisfied' level. If the top two boxes were combined, as is the common industry practice, FEDEX would have achieved a 99.6% satisfaction level.) Results are reported to senior management on a quarterly basis.

139

2. Targeted Customer Satisfaction Studies – FEDEX conducts a total of ten targeted studies on a semi-annual basis. Each direct mail survey is designed to measure the satisfaction of customers who, within the previous three months, had an experience with one of ten specific FEDEX processes, including complaint handling, claims handling, Saturday delivery, and invoice adjustments.

   FEDEX goes so far as to survey customers on their satisfaction with its credit-collection handling process. It targets the studies because a random sampling of customers would not include a statistically valid sample of those that had recent experience with the process in question. Response rates range from 8 to 18%.

3. FEDEX Center Comment Cards – Each FEDEX store-front business service center invites customer reaction via comment cards. Staff collect and tabulates the cards twice a year and forward the results to the management in charge of business service centers.

4. Customer Automation Studies – FEDEX equips roughly 10 000 of its largest customers with Powership shipping and billing computer systems. These systems allow customers to print airbills and shipping labels automatically, track and confirm package delivery, generate shipping volume and shipping management reports, and receive invoices electronically. Powership users account for about 30% of the company's total package volume. The annual direct mail survey, designed to gauge satisfaction with the automated devices, has generated a response rate of about 45%.

5. Canadian Customer Study – Outside of the United States, Canada represents the largest source and destination of Federal Express shipments. These customers are surveyed once a year, by phone and direct mail, in the same manner as used in the US Customer Satisfaction Study.

The Federal Express story is detailed in the excellent American Management Association book, *Blueprints for Service Quality – The Federal Express Approach* [4].

140

FRONTLINE PERSONNEL EMPOWERMENT FOR CUSTOMER SATISFACTION

Federal Express

The empowerment of frontline, customer contact employees for achieving Customer Satisfaction represents the second highest point total in Category Seven (Customer Satisfaction) in the Malcolm Baldridge National Quality Award. The developers and proponents of this prestigious award recognized the incredible impact these individuals can have in customer 'Moments of Truth'.

Federal Express chairman and CEO Frederick W. Smith has stated, 'Customer Satisfaction begins with employee satisfaction'. Employees are the primary focus of the company's people, service, profit philosophy for which quantifiable measures are set each year and bonuses depend on whether these objectives are achieved.

In this regard, FEDEX goes to extraordinary lengths to make their employees an integral focus of their entire operation. This is particularly true with its 40 000 frontline, customer contact employees who really represent FEDEX to its customers in over 1.5 million shipment transactions each day. At this volume, 99% customer satisfaction translates to 15 000 'perturbed' customers each day! No wonder Federal Express is never satisfied!

Beyond clearly and consistently communicating to all employees the company's two objectives of 100% Customer Satisfaction and 100% service performance on every package handled, FEDEX management believes that employees will perform beyond expectations if they know what is expected of them and are properly trained. FEDEX believes that training is a key element to empowerment so that employees are dealing from a position of knowledge in all situations rather than from one of uncertainty.

Training efforts are decentralized to the division level. A $64 million interactive video system as well as online computer terminals are two vehicles that are broadly used for delivering skills training courses. FEDEX makes a special point of providing extensive job training to all new recruits in the frontline, customer contact position.

For example, customer service agents who will deal with customers on the phone complete a five-week course before they handle their first solo customer call. Subsequently, they have four hours of job training each month and must pass a job-knowledge and skills test given via interactive video twice each year. Similarly, another category of frontline employee, couriers, receive twice-yearly recurrent training on customer service and job-performance procedures.

FEDEX designs these training and testing procedures to be reinforcing and supportive, not selective and punitive. For example, twice a year couriers have to pass a 100-question test at the computer terminal within a two-hour time limit. If the score is inadequate, they take remedial training. Even those that 'pass' receive a printout that details areas of strength and areas where they need to 'brush up'. Importantly, this type of training and testing is used for all levels of employees, including management.

FEDEX management encourages frontline empowerment through both its overall goals and attitudes. CEO Smith described these attitudes:

> There are lots of examples on a daily basis where our employee has the ability to satisfy a customer, but only at the risk of doing so either outside of policy, or where no policy exists. We want our employees to satisfy that customer's need because they feel they have the authority or the power – the empowerment – to do whatever is necessary. The employee has to feel that he or she has the right, the authorization, or the backing to do whatever is necessary to satisfy the customer.

A clear example of this empowerment is reflected in the FEDEX Billing Center where non-management employees are authorized to resolve customer billing problems without management approval for up to a $2000 credit or refund. Similarly, couriers are authorized to spend up to $250 without authorizaiton to solve customer problems. ($250 represents ten times the average $25 charge for a FEDEX package.) [4]

CUSTOMER SATISFACTION GUARANTEE

L.L. Bean, Harry and David, Hampton Inns

L.L. Bean and Harry and David are both mail-order catalogue organizations that have been a successful part of America's and International shopping practices for over 60 years. Both organizations were focused on Customer Satisfaction by their founders and survived the Great Depression by intensifying this customer focus. In an era when many major organizations are having trouble sustaining themselves through normal cyclical economic downturns, it is remarkable to study two organizations that survived *the* economic downturn and are leaders in their respective categories today by building on the characteristics that were integral to their surviving the 1930s.

Importantly, neither L.L. Bean nor Harry and David sells products that cannot be purchased from other similar organizations at comparable or lower prices. Further, both compete with sophisticated retail operations across the country. These retail organizations competing with L.L. Bean provide the customer with the opportunity to try on the garments or touch and test the equipment. In the case of retail outlets competing with Harry and David, the customer has the opportunity to squeeze the fruit.

Yet, both L.L. Bean and Harry and David leave their retail competitors in the dust, as satisfied customers return to their catalogues again and again. In fact, both organizations have taken significant business from their retail counterparts, as increasingly busy individuals find it easier and more convenient to order products for themselves – and gifts for others – from Bean's or Harry and David's catalogues and let those organizations do the wrapping and shipping. Busy individuals will pay a premium when those services are executed in the flawless manner that characterizes both these firms. The 'bottom line' is that both organizations have been able to continually grow and prosper based

143

upon their unswerving commitment to exceeding customer's expectations in both products and services.

Additionally, both L.L. Bean and Harry and David back these superior products and services with Customer Satisfaction guarantees that are single minded in their commitment to World Class Customer Satisfaction. These guarantees are presented in their entirety below:

L.L. BEAN

Our products are guaranteed to give 100% satisfaction in every way. Return anything purchased from us at any time if it proves otherwise. We will replace it, refund your purchase price or credit your credit card, as you wish. We do not want you to have anything from L.L. Bean that is not completely satisfactory.

HARRY AND DAVID

Satisfied customers are made – one at a time – through exceptional levels of quality and the 'personal touch' you'll find only at Harry and David. We want to keep you a satisfied customer. You and those who receive your gifts must be delighted or we'll make it right with either a replacement gift or a full refund – whichever you think best.

In support of their guarantee, L.L. Bean has Customer Service personnel available 24 hours a day, 365 days a year. It has often been said that residents of the brusque and impersonal northeastern section of the United States will call the Bean operators and order merchandise just for the opportunity to speak to a congenial and polite individual and one who will call them by name throughout the transaction!

While the Ritz-Carlton Hotels have received well justified accolades based on their garnering the 1992 Malcolm Baldrige National Quality Award, a chain of 250 discount hotels in the United States has set the standard in the industry for Customer Satisfaction guarantees. It can be argued that the Ritz-Carlton Hotels compete is a pretty rarefied atmosphere. Clearly, they have major competitors, but the absolute number of hotels competing at this level is relatively limited. However, the same is definitely not

144

the case among the rapidly growing number of 'no-frills' hotel chains.

These hotels are competing for the business traveler's dollars in an environment ever more circumspect of business expenses. This situation is even more acute with the advent of President Clinton's tax policies that will limit the deductibility of business expenses to 50% of the total, versus the previous deductibility level of 80%. This significant tax code change impacts not only hotels and restaurants in the United States, but internationally as well.

The Hampton Inn chain has staked their claim to an important share of this market through a direct appeal to customers and customer satisfaction. Their dedication to the customer led to their recognition by the Consumers' Union, publishers of *Consumer's Reports*, as the best of the discount hotel chains. The Hampton Inns' Customer Satisfaction guarantee is unsurpassed in the hotel industry:

HAMPTON INNS

We guarantee high quality accommodations, friendly, and efficient service and clean, comfortable surroundings. If you are not completely satisfied, we don't expect you to pay.

Our organization may not be in the direct mail merchandise business and we may not be running a hotel chain, however, if we believe in what we are doing, why can't our organization have these types of guarantees and stand behind them. Evidence shows that organizations that have this type of customer commitment and stand behind it rarely have to hide behind their reputations.

SUMMARY LESSONS LEARNED

1. Our organization must adopt Customer Satisfaction as a strategic approach and weapon. Toyota, Lexus, and Federal Express have made Customer Satisfaction their strategic focus to achieve their business objectives. They have successfully used the strategic

weapon of Customer Satisfaction to continuously improve their marketplace position.

2. Our organization needs to be structured with the customer in mind, not as an afterthought. Toyota and Lexus have designed their organizations so that the 'Voice of the Customer' is heard throughout the organization. We need to design our organization with the same purpose.

3. Customer Satisfaction can be quantified and needs to be monitored continuously to keep abreast of the ever-changing marketplace. Federal Express has developed an extensive yet easily understood measurement system for its numerous customer classifications. Our organization needs to develop an approach to quantifying customers' needs and expectations and their perceptions of our performance on these issues, as well as those of our competitors.

4. Our organization needs to clearly understand our customers' needs and expectations if we are able to succeed in 'delighting' them. Federal Express has been able to identify the needs and expectations for its various customer classifications and has also been able to rank and weight these needs and expectations. This level of information provides Federal Express with exceptional marketing power and strategic focus. Our organization needs to emulate this same approach and we will garner the same strategic power.

5. Our organization must develop an approach for empowering our employees to satisfy our customers efficiently and effectively. Our frontline customer contact personnel are both our eyes and ears to our customers. Moreover, they invariably transmit our organization's first and most frequent impression to our customers. Federal Express has learned that satisfied customers derive from satisfied employees and empowering these employees is an important component to their job satisfaction.

6. Our Customer Satisfaction guarantees are only worth what we want to place behind them in terms of organizational support. L.L. Bean, Harry and David, and Hampton Inns have staked their entire organizations to their 'no nonsense' Customer Satisfaction guarantees. Our organization needs to determine the

146

strongest possible guarantee it can make and then to stand stead-
fastly behind that guarantee.

REFERENCES

[1] *Blueprints for Service Quality - the Federal Express Approach*,
     American Management Association, New York, (1991), p. 16.
[2] *Maximizing Customer Satisfaction: Meeting the Demands of The New
     Global Marketplace*, Business International, New York, (1990) pp.
     41–2.
[3] *Maximizing Customer Satisfaction*, December 1990, Toyota case study,
     pp. 41–49, the Economist Intelligence Unit.
[4] *Blueprints for Service Quality - The Federal Express Approach*,
     American Management Association, New York, (1991), pp. 55–64.

An approach to customer satisfaction research

6

You just listen to the customers, then act on what they tell you.

Charles Lazarus, CEO Toys-R-Us

The Customer Satisfaction Best Practices presented in Chapter 5 are all premised on an in-depth understanding of customer needs and expectations. While Federal Express may be our World Class example of the best customer expectation identification process, all the organizations included in our Best Practices pantheon regularly conduct customer research on ever-changing needs and expectations and on the all-important subject of customer's perceptions of their performance.

The objective of this chapter is to recommend a process for our own Customer Satisfaction research. Clearly, the approach presented is not the only way to address this subject, but it is consistent with our overall direction to Customer Satisfaction – it is based on common sense and it endeavors to get our organization the maximum impact with the minimum of expense and exertion. This does not mean that creative thinking and hard work are not required in conducting any level of Customer Satisfaction research. They certainly are. However, our objective is to be as efficient and effective in our Customer Satisfaction research approach as we want our organization to be in initiating our Customer Satisfaction effort.

RESEARCH BEGINS INSIDE AND OUTSIDE OUR ORGANIZATION

In Chapter 4 we stated that we need *not* initiate our Customer Satisfaction research effort by enlisting the services of professional research suppliers. Some preliminary information on customer needs and expectations can be developed by conducting research about our customers *inside* our organization. In this chapter we will present an approach to this type of research with examples from several different organizations that began their research journey along this same path. A great deal can be learned about our customers by 'asking the experts' inside our organization. The best part about this type of research is that we have basically already paid for this information by having these 'experts' on our payroll and they are probably dying to share their knowledge with us!

We will also discuss an approach to traditional customer satisfaction research conducted outside our organization. We will examine the areas this research should encompass, the types of issues that may arise when dealing with professional research organizations, and we have even included some prototypical presentations made by World Class Customer Satisfaction research firms so that we might know what to expect from preliminary presentations from professional firms.

Finally, we have also included some research questionnaires in the Appendix for informational purposes only. They are included to demonstrate what Customer Satisfaction research questionnaires *can* look like. Clearly, they are *not* presented as the definitive research vehicles, nor as a 'one size fits all' approach to questionnaire development. Rather, they are included to take the mystery out of a subject that has been needlessly complicated and clouded by some professional practitioners who believe the best way to justify high fees is to complicate and obfuscate. Our goal, on the other hand, is to clarify and simplify.

LOOKING INSIDE FOR OUR EXTERNAL CUSTOMER

Our customer satisfaction research efforts inside our organization can include obtaining valuable information in three basic areas:

1. customer needs and expecations as currently understood by our frontline customer contact personnel; 2. actual statistical information about our performance regarding these needs and expectations; and 3. preliminary competitive information to include both the identification of competitors and their current practices and performance as it impacts our customers. Even a modicum of information from each of these three subject areas will provide us with a substantial advantage over our previous 'data-free' state.

AN INSIDE PERSPECTIVE ON CUSTOMERS

As we learned in Chapter 3, our organization impacts the customer far more than we may previously have believed. These customer contact 'Moments of Truth' are occurring almost continuously and now is the time to begin gaining some precious information from these interactions.

We should already have developed a rough estimate of our organization's 'Moments of Truth' – where are they occurring, how often, who from our organization is involved, and what basic information is being exchanged during these 'touches' with the customer? Again, if we are just beginning this journey, even an approximate count is sufficient at this juncture. Once we have that, we need to prioritize which customers, or group of customers, we are going to study first. There is no 'school book' answer to this issue. We, as an organization, need to determine our customer priority. Is it to be based on historical performance, current levels of activity, potential levels, business travelers, vacation travelers, hip-replacement patients, heart patients, etc.

Once we have prioritized our customers for our research purposes, we need to do the same for those who contact or touch them. Which group is most frequently in touch with our target customers? Do we have actual data to support this priority or are we dealing with 'warm feelings' on the subject? By this time in our journey we should have a strong predilection for hard, quantifiable data and it will be even more useful and necessary as we begin to tighten our Customer Satisfaction research focus. It is not important what we 'think' about who really comes in direct contact with our customers. What is important is what we 'know'.

A typical example of the disparity between 'thinking' and 'knowing' occurs in many hospitals. When asked about who has the most frequent contact with their patient-customers, many hospital administrations will answer doctors and nurses. In fact, they may be correct about nurses, but most studies on the subject demonstrate that house-keeping and food service personnel have far more frequent and regular contact with the vast majority of patients than do doctors. At least the patient is conscious when they are served a tray, which may not be the case when the doctor comes in contact with them.

FOCUS ON THE CUSTOMER THROUGH FOCUS GROUPS

Once we have focused our research spotlight on the appropriate internal contact, we need to begin the process of transferring their understanding and knowledge of the customer to our strategic customer satisfaction process. An impactful and efficient means of doing this is through 'focus groups'. These small, informal sessions can be 'gold mines' of information if properly structured and conducted.

Here are a few simple tips for conducting successful focus groups:

1. Keep the groups small – no more than five or six individuals. Professionally moderated groups are usually slightly larger, but we are conducting this research ourselves, so let's make it easy at the outset. This smaller number of participants allows everyone to participate and also it does not let anyong 'hide' by not participating, which is always possible in larger groups.
2. Target the sessions by clearly stating the objective at the outset, printing it on an easel or blackboard if necessary. This will help keep the group 'focused' and provide a rein should discussion begin to wander off-target – into 'solving' problems, for example. In general, participants will be open and honest if they believe their input will be used and useful and that their comments are being made without attribution or fear of retribution.

152

3. Make the sessions open and friendly. This can be done through conducting them in some quiet area of our facility, preferably one 'off the beaten path' to avoid distractions and calling unnecessary attention to our effort. Make the sessions informal, possibly by casual dress requirements or serving refreshments, lunch, or dinner afterwards. Certainly the presence – or promise– of food will positively impact attendance!

4. Be an 'active listener'. We are there to listen, not participate. Participants may say things we disagree with or do not believe, but the purpose of these groups is to 'debrief' our front-line customer contacts, so we want to hear what they have to say. A tremendous help in this regard is to record the groups on audio tape. This provides the opportunity for review at a later date and keeps our attention on the discussion rather than on feverish note taking. Participants will quickly forget the recorder's presence as they get caught up in the discussion.

5. Have a discussion guide and make the topic list manageable. Do not try to capture the entire organization's bounty of knowledge in one sitting – a whale is not eaten in one bite. Rather, have a specific and predesignated list of subject areas that can be covered in about one hour's discussion. These subject areas will vary greatly by organization and we should spend time in advance of these sessions determining at least a preliminary list of subject areas. In all probability, these groups will generate additional areas we want to discuss, possibly of an even greater priority than those on our initial list. That is the beauty of this type of initial, informal, and internal research.

Here are some subject areas which many organizations have used in these types of initial focus group discussions. We want to learn from our organization what our customers want regarding each of these items. We will build upon this list when we conduct external Customer Satisfaction research.

Clearly, this list is provided only to spur our own subject creating process, so it should not seen as all-inclusive or as an absolute. Additionally, whenever we see or use the word 'product' we should always consider the word 'service'. As we have

153

discussed throughout this book, we are all in the service business.

POSSIBLE FOCUS GROUP DISCUSSION AREAS

Service or product characteristics
Performance
Price
Presentation/Packaging
Promotion/Publicity/Advertising
Service or product reliability
Durability
Downtime
Customer Service Levels
'User friendliness'
Ease of Customeer Contact and Assistance
Frequency of repair or service request
Most frequently requested assistance or question
Organizational issues
Empowerment of customer contact personnel
Level of training, depth of knowledge required or preferred
Frequency of contact preferred by customer
Response time to customers
Financial issues
Payment terms
Ease of payment
Guarantees, warranties, return policy
Payment/billing dispute process
Competitive issues
Known performance levels
Customer statements on performance
Customer comparative perceptions

Clearly, these are just a sampling of possible topics for discussion in our focus groups. In each case, we are trying to determine from our internal sources what our customers are expecting from our organization. Frankly, we are not interested in what our views on these subjects are, nor should these groups degenerate into a discussion by participants concerning what they believe our performance should be in each. Rather, we are listening for the 'Voice of the Customer' through those in our organization who come into the most frequent contact with them. Invariably, that voice of the customer will come out in these discussions, as our frontline personnel share with us the comments, compliments, complaints, and conundrums customers have expressed to them.

The results of these internal qualitative studies may be sufficient for smaller organizations to begin prioritizing their action plans aimed at better serving our customers. For larger organizations, they can form the basis for further quantitative research which we will examine later in this chapter. In either case the results are invaluable to our organization because they represent real learning about our customers.

Importantly, this qualitative first step is always useful. It opens discussion within our organization on the subject of the customer and the primacy of Customer Satisfaction and that is always a positive step. Additionally, it facilitates the communication between various parts of the organization and between the body of the organization and management. This is also an important benefit, even if our organization currently enjoys an open and positive internal atmosphere. Finally, every World Class Customer Satisfaction research organization includes in their preliminary research steps this aspect of qualitative organizational debriefing of individuals inside our organization. They will focus these research efforts on customer contact personnel, at a minimum. So our endeavors in this area will only facilitate our quantitative research efforts if we decide to pursue that course some time in the future.

ACTUAL ORGANIZATIONAL PERFORMANCE LEVELS

One of the key elements of our *Making Customer Satisfaction Happen* Model that we discussed in Chapter 4 was the need to determine actual organizational performance levels. Specifically, the third step of that model deals with measuring customer perceptions of our performance on a variety of attributes that have already been prioritized in importance by our customers. Once we know how customers perceive our organization – and their perception is our reality – we need to determine how well, or poorly, we actually perform these various tasks.

This information can be gathered quickly and easily from our internal organizational measurements. As an organization we had better be measuring our key processes and measuring them from the perspective of our customers. If we have not been conducting our organization measures from that customer perspective prior to this time, we had better bring our measurement effort 'on course' and do so quickly.

The entire subject of effective organizational measures is discussed in detail in *Making Quality Happen*. However, the bottom-line point on measurement is that, at a minimum, our organization should be measuring the outputs of its primary processes dealing with the customer. We then need to compare this measure of actual performance to the customer's perception of our performance. We also need to make sure our measurements are calibrated from the customer's perspective.

More often than not, they are not. Therefore, we invite the type of customer arguments concerning topics such as 'two day' delivery that we previously mentioned. Our organization believes the 'two days' starts when the package leaves our loading dock, the customer believes the 'two days' starts when they hang up the phone from placing the order. If there is ever any question as to whose view is the 'correct' one, just consider who is paying the bill!

If we know how we are perceived by our customers on specific products or services and we know what our actual level of performance is, we can concentrate on improving that level of performance and utilizing measurement of that process as our measure of improvement. It is not necessary to continually recheck with our customer concerning our progress. Clearly, this continual, or 'real time' customer assessment of our processes would be ideal, but it is clearly not economically or technologically feasible for the vast majority of organizations.

However, we must make certain that we do not lose sight of the fact that customer's needs and expectations are ever-changing, so the level of performance they may have been seeking today may not be what they are looking for tomorrow. The marketplace can change rapidly – the shifting customer paradigms – and our one time organizational strengths may no longer be required or even relevant. Consider the impact that Lexus had on almost every aspect of the luxury car market. It will never be the same, so there is no need to measure 'ancient history'. Therefore, we need to build into our process some customer contact on a regular basis, as the former mayor of New York City Edward Koch was wont to say, 'How Am I Doin'?' We need to ask our customers that question as frequently as possible.

COMPETITIVE INFORMATION

Just as our organization should know its actual performance level on all processes important to our customers, we should also know how our competitors are performing on these same processes. As we learned in Chapter 2, our service failures are far and away the leading reason for our losing customers to the competition. This fact is depressing enough, unless we do not even know that it is occurring, which is exactly what is happening if we do not know our competitor's level of performance.

We can obtain this competitive information from a variety of sources. Certainly the first place to look is within our own organization. Our customer contact personnel may be very familiar with the comparative processes and performance levels of our direct competitors. They may be hearing about them from our shared customers on a daily basis. So our own internal sources are an excellent place to start. This competitive performance topic is one ideally covered in our focus groups.

Importantly, it may be appropriate to make this 'competitive analysis' the topic of entirely separate focus group sessions *after* we have solidified our understanding of our customers' perceptions of our organization. These competitive performance focus groups could follow a similar discussion guide to that used for our internal organizational analysis. This will aid in a comparison between our organization and those who are looking to take away our franchise.

Next, we can conduct our own 'sampling' of our competitor's performance. In Chapter 4 we discussed how Sam Walton would not only try to visit as many of his own WalMart stores as possible each year, but how he would also make time on each of these visits to 'shop' the competition and recommend 'on the spot' competitive merchandising counterattacks when necessary!

Customer Service benchmark L.L. Bean does that same type of competitive 'shopping' by ordering products comparable to theirs from competitive catalogues. Bean compares not only the product received but also the all important service features of the entire transaction. Bean's research has demonstrated that it is these service features that are highly valued by the catalogue shopping customer and therefore Bean excels in them and keeps a wary and comparative eye on their competitor's performance.

An additional rich source of information concerning our competitors can be our suppliers. They may be very familiar with the operations and practices of some or all of our competitors. While our suppliers may not also be serving our competitors, they may well have industry contacts that can provide extensive information about very basic competitive strengths and weaknesses. These may include such subjects as cycle times for manufacturing, order fulfillment, product development, and customer service. Also they may be a source for learning more about customer guarantees offered by our competitors or new product or service offerings under consideration.

This type of information gathering from suppliers should not be construed as some form of 'industrial espionage'. That is clearly not our purpose. Rather, our objective is to make our supplier a more integral part of our organization, as if they were part of its internal operation. Stunningly, a large number of organizations hold their suppliers at arm's length and never integrate them into the operation of their processes. In many cases these short-sighted organizations thereby lose an extensive source of industry and competitive information.

When one organization finally approached one of its suppliers about creating more of a 'partnership', the organization was shocked at the depth of knowledge the supplier had about the industry, the competitive factors facing the industry, and the technological advances that would be necessary to succeed in the future in this industry. The organization's management sheepishly asked

158

the supplier's management how they had such a vast source of information.

Their reply was quite simple and logical. Their offerings as suppliers to this major organization constituted only a minor part of one of the organization's product lines. However, this product category represented the supplier's 'life blood'. While the loss of this product for the customer organization would hardly have been noticed at their headquarters, it would have had serious – if not fatal – implications to the supplier's organization. It was incumbent upon the supplier to know everything they could about every facet of the industry. Therefore, it is appropriate that we include our suppliers in this customer 'debriefing' just as we are including our own organization's customer contact personnel. We could be just as surprised and delighted at the results.

MOVING FROM INTERNAL TO EXTERNAL

For some organizations, this initial level of internal Customer Satisfaction research is all that is necessary. In many cases, the level of information gained from this first phase of research will provide sufficient Customer Satisfaction information to accurately target the organization on the first steps of its Customer Satisfaction journey .

Evolving from this internal qualitative Customer Satisfaction research is external quantitative customer research. Clearly, the objective of *Making Customer Satisfaction Happen* is *not* to make its readers experts in the field of Customer Satisfaction research. There are World Class organizations that do this type of research and are eminently qualified to conduct studies that are appropriate for almost any type of organization.

However, it is our objective to make our readers 'informed consumers' who know exactly what they are buying and what they should be expecting to receive for their investment. The major content of this discussion is based upon detailed interviews and discussions with seven of the foremost Customer Satisfaction research organizations. While each of these organizations has their own special technical arsenal of research tools, their overall approach to the subject of Customer Satisfaction research is consistent and exceptionally logical and that is what will form the basis of our discussion on this subject.

BASIC RESEARCH APPROACH

Any Customer Satisfaction research project – and any organizational project for that matter – should follow the basic structural outline of having an *objective*, a *strategy* for achieving that objective, and a set of *tactics* that execute the strategy. This may sound like an organizational primer, but often organizations jump directly into the tactics – the design details – of Customer Satisfaction research before understanding why they are conducting the research, who the study will be conducted upon, what we expect to learn from the study, and what actions we plan to take as a consequence of the research.

RESEARCH OBJECTIVE

Our Customer Satisfaction research should have a clear and concise objective. Time spent developing this objective will eliminate wasted time when the actual research suppliers are involved and the 'taxi meter' is running. Focus is particularly important in our objective statement, for it is here that we are defining the 'whats' and 'who' of our research. Developing the objective statement is the responsibility of our organization. Those whom we contract with to actually conduct the study can certainly advise us on the objective, but we must 'own' this statement and must control the content of it, as the objective directly impacts what the research will be designed to provide.

Therefore, our objective should detail the basic knowledge area we are seeking to understand and the target customer from whom we are seeking to learn this information. As you can see, the preliminary qualitative research we conducted internally is a tremendous asset at this juncture, as it already provides our organization with an agreed upon focus. (Or if we found from our own internal research that we did not cover the proper target, now we can adjust our focus to the appropriate research target audience.)

160

It is imperative that our objective statement include a specific and detailed description of the target audience for our Customer Satisfaction research. The more specific and defined our research target is, the more actionable our results will be. Additionally, this detailed customer description will also help control research costs, as the supplier will have a clear target for their screening efforts.

Objective statements should be brief and focused. Examples of Customer Satisfaction research objectives from several organizations include:

Research will determine the relative needs and expectations of the medical supply distributor concerning medical products suppliers and their comparative performance versus these needs and expectations.

Research will delineate requirements of orthopaedic surgeons concerning surgical instrumentation used in hip replacement operations and the comparative performance of currently marketed instruments.

Research will determine the relative needs and expectations of the business traveler concerning hotel stays of 1–2 nights at a variety of price ranges and the perceived delivery of those needs and expectations by current industry providers.

An action step is also an important ingredient in the basic objective statement of a Customer Satisfaction research project. This action step should clearly delineate exactly what actions the organization is going to take based upon the results of this research. For example, is the strategic plan of the organization going to be impacted by the research findings, is our marketing plan going to reflect our learning from this study, are our organization's annual priorities going to be restructured based upon what we learn from talking to our customers? Clearly, each of these elements should be directly impacted by our customer satisfaction research, but we need to delineate that clearly and completely in an action step statement.

The inclusion of an action step statement in the objective of our research avoids the all too frequent occurrence of the research results arriving and the organization then not knowing what to do with the findings. In that instance all the organization does know is that they have spent large sums of money without preplanning their next steps. Organizations are often good at spending money, particularly large sums of money on something that sounds as appropriate, timely, and fashionable as Customer Satisfaction research. What they often fail to do well is advance planning. Including an action step in the objective statement of the research design helps avoid that problem.

RESEARCH STRATEGY

Our research strategy should also be developed in advance and in conjunction with the organization that will actually be conducting the research. Our strategy focuses on identifying our key target segments, their relative priority, areas of information to be covered within each segment, and whom from our customer organizations will be covered in each part of the study. Additionally, we should detail what competitive aspects we want to examine – for example what specific competitors will we ask our customers about?

We need to proactively identify these items in our research strategy. Again, our preliminary qualitative research should have provided some, or all, of this information, so this strategic work may already be completed when we get to the larger scale quantitative studies.

Our organization should play the major role in developing our research strategy. Clearly, professional research firms can provide some important input into the process, but by this time our organization should be capable of identifying our priority customer targets, our competitive information needs from these customers, and the basic Customer Satisfaction issues foremost in the minds of our customers. These are the basic elements of our research strategy.

RESEARCH TACTICS

Our organization has primary responsibility for developing the research objective and strategy statements. At this juncture many might ask just what the professional research suppliers will provide.

Their expertise flows from our objective and strategic development. The research suppliers we select will develop and deliver the research tactics – the 'nuts and bolts' of the actual Customer Satisfaction study. Included in this very important phase is the development of the actual research questionnaire, the choice of how the questionnaire will be administered – mailed survey, telephone study, personal interview, shopping center intercept, etc. – and the tabulation of the final results.

Clearly, our organization should play an important role in every aspect of the research tactics, but we should not feel that we must be questionnaire development and design experts, nor should we feel – or be made to feel – that we must choose the appropriate research vehicle. This is what we are paying the research supplier for. This is their area of expertise. We should have direct and final input into this decision because we are paying for the study, but we should never feel, or believe, that it is incumbent on our organization to develop or execute these tactical issues.

ROLE OF THE PROFESSIONAL RESEARCH SUPPLIER

This is an appropriate juncture to cover some important elements to consider in dealing with professional research organizations. The Customer Satisfaction research organizations presented in this book do not comprise the entire universe for this type of research. However, they do represent a sample of World Class Customer Satisfaction research firms that have had extensive experience with this particular and unique form of research. They will not be learning on our time.

Customer satisfaction research experience

Therefore, the first step to consider when interviewing research firms other than those recommended in this book is to determine their actual depth of experience in the area of Customer Satisfaction research. Many market research firms believe that they can easily segué into this growing field based solely upon their experience conducting traditional market research. Their rationale is that 'research is research' and one study is no different from another.

163

While Customer Satisfaction and traditional market research have numerous similarities, our objective is to have a research firm that is a partner in this process, not merely someone who develops and executes questionnaires. Therefore, we should ask any potential research supplier to identify Customer Satisfaction Research studies they have conducted and provide references from customer firms whom we may choose to contact. Any hesitancy on the part of suppliers to provide this information should cause our organization to seriously question their capabilities to execute our customer satisfaction research needs.

Presentation of research findings

Some other 'needs and expectations' that we should discuss with potential suppliers include their ability to present research findings, make specific action recommendations based upon the research results, and provide organization orientations on the subject of Customer Satisfaction research.

Any organization we consider as our potential partner in this Customer Satisfaction research venture should have the ability and the desire to present the actual research findings to our organization, and our management in particular. This may sound like a 'given', but many so-called Customer Satisfaction research suppliers merely do a 'data drop' wherein they deliver reams of tabulated research results to the customer organization and quickly exit stage left, leaving us with the unenviable task of having to wade through these data and develop some conclusions.

Clearly, any research organization we consider should not only tabulate the research results, but also present these results in a manner that leads to marketplace action recommendations, rather than the need for further 'cross tabulations' and re-runs of the data.

Research results should be presented in clear and concise 'management summaries' – featuring easy to understand graphs and small words – and slightly more detailed summaries for those parts of the organization charged with actually carrying out the action steps developed as part of our original research objective.

Additionally, the supplier we select should want to personally present these data to our management. Clearly, we want to understand what will be presented and in what format, but our supplier should feel a sense of ownership and partnership so that they want to share their important learning with our organization and our management in particular. This presentation by the supplier also reduces our processing 'cycle time', in that it eliminates the need to refer back to the supplier questions that may arise in management reviews. The supplier will be able to answer these questions immediately if they are presenting the research findings.

Specific action recommendations

An important part of these research findings and another key 'need and expectation' we should have of our Customer Satisfaction research supplier is the development and presentation of distinct and clearly defined action steps based upon the research findings. These action steps should be related to the action steps included in our research objective, but they may include other recommendations based upon specific results generated by the research.

Quite frankly, one of the items we should be looking for in a research partner is their 'brain power' and we should want to and be able to tap into that source, particularly regarding actions we consider taking based upon this research.

For example, one action step our organization might have identified in our research objective was that advertising strategies would be reviewed to reprioritize key 'selling messages' based upon the results of the latest Customer Satisfaction research. We should look to our supplier partner to recommend what they believe the specific reprioritization should be – 'we should emphasize package traceability in advertising, as that expectation was in the customer's top three based upon this latest research'.

Orientations on customer satisfaction

Another element that we should look for in our Customer Satisfaction research partner is the ability and willingness to provide orientations on the subject of Customer Satisfaction. Clearly, this item falls into the 'nice to have' rather than 'need to have' category concerning potential suppliers. However, it is our experience that the World Class Customer Satisfaction research suppliers are delighted to present orientations on the subject both to our internal organization and as a 'value-added' service to organizations that are our customers.

The truly expert suppliers are totally conversant with the subject of Customer Satisfaction research, are comfortable with its nuances, and have the experience to present this information in a manner that transcends differences between various organizations. The sophisticated practitioner realizes that the more individuals within our organization that understand the benefits of this type of research, the more likely that the results will be used throughout the organization, and therefore the more likely we will be to continue this journey.

Moreover, the World Class research organizations know that every opportunity they have to interface with our customer organizations provides them with both an opportunity to better understand this customer on our behalf and the chance to possibly convert this customer organization to consider some level of Customer Satisfaction research of their own.

RESEARCH SUPPLIER PRESENTATION

Now that we have an understanding of what we should be looking for in a Customer Satisfaction research supplier-partner, what might their exploratory presentation to us look like?

To this end, we have included an actual Customer Satisfaction research model used by one of the World Class suppliers presented in the Appendix following this chapter. Clearly, this is a generic approach which we should expect to have customized to our specific organizational needs by potential suppliers. However, this model is useful for illustrative purposes to understand how one major practitioner in the field approaches the subject with potential clients (Figure 6.1).

Objectives

This initial step is completely consistent with our recommended approach to developing an overall objective for this research study.

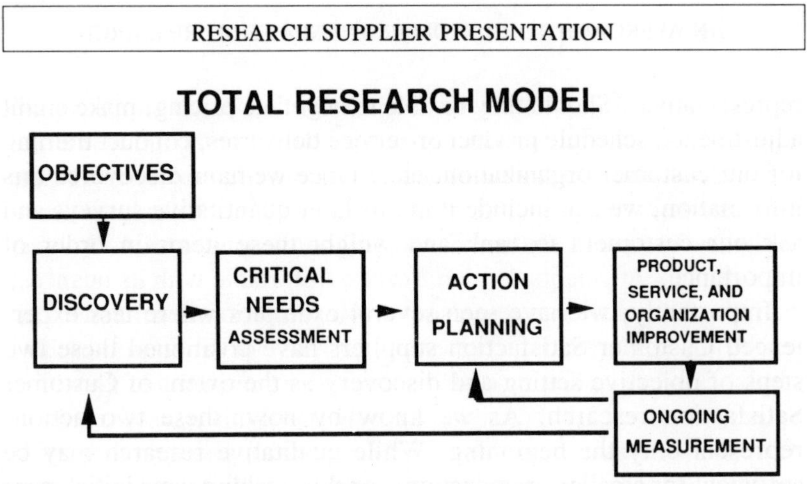

TOTAL RESEARCH MODEL

Figure 6.1 The Total Research model: an approach to Customer Satisfaction research.

Total Research believes that the input from all elements of the organization is important in developing a clear and concise research objective that can then be delivered by the research supplier. An analogy to felling a tree is often used to describe this initial step of developing objectives for the study. Specifically, if we were given eight hours to complete the task of felling the tree, then we would be best served to spend six of those eight hours sharpening the axe!

Discovery

This is a qualitative research stage in which we determine customers' needs and expectations from both our internal organization's perspective and from that of our external customer. An important aspect of this step is to determine the customer's needs and expectations in *their* language – how do they describe what they want and need? By determining how our customers articulate their needs and expectations, we are better prepared to conduct actionable research on these subjects in a manner and style that our customers are comfortable with.

An example from one study of the type of information that may surface in this Discovery phase is the importance to our customers of our field representatives. In this discovery phase we want to understand just what our customers are expecting from our field

167

representatives. Should they be able to negotiate pricing, make credit adjustments, schedule product or service deliveries, conduct training for our customer organization, etc.? Once we have discovered this information, we can include it in our later quantitative surveys and ask our customers to rank and weight these items in order of importance.

Importantly, we have seen several examples where less experienced Customer Satisfaction suppliers have positioned these two steps of objective setting and discovery as the extent of Customer Satisfaction research. As we know by now, these two actions represent only the beginning. While qualitative research may be sufficient for smaller organizations, or those taking their initial steps on this journey, focus group discussions are never a long term substitute for quantitative, projectable research conducted with a significant portion of our customer base.

Critical needs assessment

The first two steps are now called into play in critical needs assessment. This step comprises the actual quantitative study of Customer Satisfaction with our organization – and our competitors – on the attributes our customers have identified in preliminary qualitative studies. The 'deliverable', at minimum, of this step should be a listing of customer needs and expectations that is ranked, weighted, and for which we have quantitative evaluations of both our organization's performance and that of our major competitors. From this information we can develop a gap analysis matrix to clearly identify strategic organizational opportunities, as we discussed in Chapter 4.

Importantly, this is not an 'all or nothing' step. We do not need to do one huge study at one time. Rather, the sophisticated and financially scrupulous organizations conduct this step in waves. They 'pilot' the quantitative study with a test audience and then expand the research based upon their learning from this pilot test. The expansion is in waves, so that refinements and improvements can be made to the research questionnaire, or research technique, as the study progresses.

While this 'rollout' approach may take more time than a 'one shot' blitz, it is far more prudent, efficient, and effective because there is always additional and incremental learning involved in every study. Even the most seasoned Customer Satisfaction research suppliers are learning something new in each study, and each wave of each study, that they can apply in future waves.

Action planning

This step involves developing a plan to address the 'gaps' illustrated in the matrix of our customer perceived performance versus their needs and expectations. Importantly, we should expect our research team to have clear recommendations for our organization concerning this action planning. This is where their analytical ability should come to the fore and provide some specific, actionable, and tactically feasible recommendations based upon the research results.

One research organization used an historical analogy to explain their strength versus their competitors in this action plan recommendation phase. The story is told of Ambassador Joseph P. Kennedy and one of his son John Fitzgerald Kennedy's numerous political campaigns. Ambassador Kennedy was omnipresent in the inner workings of the campaign and come election night he was hovering anxiously at Kennedy campaign headquarters as the votes came in. Ambassador Kennedy asked one of the campaign directors how the results were going and this individual replied, 'We're ahead by about 10 000 votes'. The Ambassador's brusque reply cut to the heart of the matter, as he bluntly stated, 'I know by about 10 000 votes, I expect you to know by *exactly* how many votes!'

Therefore, we should not expect 'warm feelings' or thoughts from our supplier, but rather clear and reasoned actions that can make a difference for our organization in the marketplace.

Ongoing measurement

The Total Research model includes this important step that is often overlooked by organizations. This step involves developing organizational measures which directly correspond to the prioritized needs and expectations identified by our customers in the research we have just concluded. Specifically, our organization should now be measuring our actual performance on the attributes our customers ranked

169

as most important. For example, if filling our customers' orders completely on the first attempt was identified as a priority need and expectation, then our organization had better establish an ongoing measure of our actual peformance on this attribute. Or maybe the customer wants their telephone calls to our organization answered within three rings, then we had better establish this measure immediately and begin tracking our performance against it.

As we discussed in Chapter 4, once we know our customers' prioritized needs and expectations, we can begin to chart our actual organizational performance against these measures. It is then not necessary for us to conduct another major study before we know if we are making improvements on these processes that our customers have told us are important to them. However, we should make certain that we develop an organizational strategy for determining Customer Satisfaction that includes regular customer studies to the extent financially feasible for our organization.

This model is just one organization's approach to the subject of Customer Satisfaction research. However, it is entirely consistent with our *Making Customer Satisfaction Happen* model described in Chapter 4 and, even more importantly, it represents a common sense and logical approach to the subject. Each of the World Class suppliers we present in the Appendix will have a different visualization of this process. However, the basic components are consistent across all these organizations. Their tactical approach to various elements may vary, but the overall objective and strategic direction are consistent. That is one of the reasons why they are World Class Customer Satisfaction research suppliers – they are following a proven approach to the practice of learning what customers want and how they perceive our organization.

SUMMARY PERSPECTIVE

The one eyed person is king in the land of the blind. Even some knowledge of our customers' needs and expectations is better than none. Actionable information is always more useful than reams of dust-gathering data. The vast majority of organizations have very limited customer knowledge. In today's global environment, no marketplace is static for long.

170

Making Customer Satisfaction Happen should be a 'wake up' call to some organizations, a call to action for others. It should alert us to gaps that may exist in our customer knowledge base and prompt us to address those issues. We have presented several scenarios in a variety of markets where major organizations believed the *status quo* would last forever. They took their customers for granted and paid an enormous price, figuratively and literally, when the paradigm shifted and competitive forces targeted the franchise they believed to be theirs by divine right.

We can do our best to make sure this fate does not befall our organization by proactively addressing the issues of *Making Customer Satisfaction Happen*. They have been consciously designed based on a common sense approach unifying our personal and professional values and standards. None of the approaches to implementing

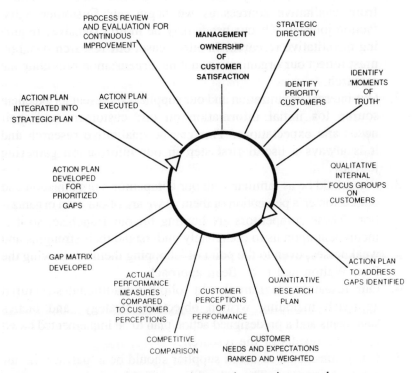

Figure 6.2 Customer Satisfaction: an integrated, strategic approach.

171

Customer Satisfaction discussed herein involve additional staffing, unreasonable financial expenditures, capital equipment or even a knowledge of higher math. Moreover, the marketplace is replete with examples of organizations in a variety of industries and disciplines that have utilized these and similar approaches for significant marketplace impact. We can begin the process today, if we so choose, because our competitor may have made that same decision yesterday!

SUMMARY LESSONS LEARNED

1. Our research program should be tailored to our organization's needs rather than be a generic or 'one size fits all' approach. Our organization may need information that can be obtained from qualitative sources as we begin our Customer Satisfaction journey. Or our needs may be more extensive, requiring quantitative research. In either case, our research program must reflect our organization, not the organization providing the research.

2. Our internal organization and our suppliers represent an excellent source for initial information on our customers and their needs and expectations. This can be qualitative research and it is always a useful first step in our information gathering journey.

3. We should be as familiar with our competitor's organization and our customer's perception of them as we are of our own organization. These competitors are hunting for our franchise, so it is incumbent upon us to completely understand their strengths and weaknesses, even to the point of 'shopping them', following the Sam Walton and L.L. Bean approach.

4. Our research program should follow a traditional structured approach including written objective, strategy, and tactics statements and a predesigned action plan to be implemented based upon the results of the research.

5. Our professional research supplier should be a 'partner' in our research journey. They should provide detailed input on our

research strategy and tactics, as well as presenting the results and recommending action steps based upon these results.

6. Learning about our customers is a continuous process. If we want to maintain our customers in the future and keep our organization viable, then we can never stop learning from our customers and learning about them.

Appendix

Two additional items have been included in this Appendix to assist in *Making Customer Satisfaction Happen* in any type of organization. The first is a listing of Customer Satisfaction research resources. The second is a generic example from part of an actual customer satisfaction research questionnaire.

CUSTOMER SATISFACTION RESEARCH RESOURCES

The seven organizations listed in this Appendix represent some of the foremost Customer Satisfaction research resources available. Clearly, they are not the only such organizations, nor does their inclusion in this Appendix represent any guarantee. However, these organizations, and their customers, have been interviewed in detail concerning their skills. Importantly, the majority have international capabilities and all have international contacts. Moreover they represent an excellent starting point for any organization commencing the journey of Customer Satisfaction.

CUSTOMER SATISFACTION RESEARCH QUESTIONNAIRE

This questionnaire is provided as an example of what a Customer Satisfaction research vehicle might look like. Clearly, 'one size does not fit all' and this format is only provided to demonstrate the depth of questions some studies have utilized. It is important to remember that questionnaire development is really the province of the Customer Satisfaction research professionals. However, more knowledge is always better than less when it comes to our customers, and so we have included this example for reference purposes.

Customer Satisfaction research resource

Expertise US [X]   International [X]

Name | WALKER: CSM

Address | 3939 PRIORITY WAY SOUTH DRIVE INDIANAPOLIS, IN. 46280-0432

Phone | 800-334-3939   Fax | 317-843-8548

Key contact | Sheree L. Marr   Title | Senior Vice President

Skill summary (techniques, models, process names)

One of the largest organizations in the field with extensive experience across sectors and internationally. Part of a larger Walker research firm which provides additional services and capabilities.

Customer Satisfaction measurement seen as an integral part of TQM, not merely a research project, or one time event, but one with cross functional and strategic implications.

Client listing for Customer Satisfaction studies

IBM, Eastman Kodak, Baxter, Eli Lilly, Procter & Gamble (Paper, Health & Beauty Aids), General Mills, Anheuser Busch, Dial, Nutrasweet, Motorola, Florida Power and Light, AT&T, Caremark, GTE.

Additional capabilities (training, presentations)

Can conduct orientations, presentations, and training on customer satisfaction, from overview to detailed half-day, tactical sessions.

CUSTOMER SATISFACTION RESEARCH RESOURCE

Expertise US [X]　　International [X]

Name

TOTAL RESEARCH CORP.

Address

5 INDEPENDENCE WAY, CN-5305 PRINCETON, NJ
08543-5305

Phone [609-520-9100]　　Fax [609-987-8839]

Key contact [Theresa A. Flanagan]　Title [Vice-president]

Skill summary (techniques, models, process names)

Total research/quality management (TR/QM) is a fully developed approach to strategic marketing. Uses customer-defined quality issues to drive performance of organization.

TR/QM is a process for measuring, managing, and improving: customer quality perceptions; your performance – and your competitors – versus these perceptions; quality image impact of various product-service variables; internal quality issues reflecting employee satisfaction.

Client listing for Customer Satisfaction studies

Motorola, Chevron Chemical, DuPont Pharmaceuticals, Vista Chemical, Alcoa, IBM, Florida Power and Light, Ryder Trucks, Spri..t, Merrill Lynch, Chemical Bank, GTE.

Additional capabilities (training, presentations)

Can conduct orientations, presentations and training on Customer Satisfaction, from overview to detailed full day, tactical sessions. Delphi group provides quality consulting capabilities including management and systems alignment, work climate alignment, process improvement and redesign, supplier quality alignment, cross-functional process improvement, performance measurement linkage.

178

Expertise US [X]   International [X]

Name [**QUALITY STRATEGIES**]

Address [7850 NORTH BELT LINE ROAD, LAS COLINAS, TEXAS, 75063]

Phone [214-506-3431]   Fax [214-506-3612]

Key contact [Corinne Maginnis]   Title [President]

Skill summary (techniques, models, process names)

Quality strategies, a M/A/R/C group company, is part of one of the largest research companies in the world. Services of the M/A/R/C group include custom market research, Customer Satisfaction measurement, marketing modeling, forecasting, simulation, and database target marketing. The Action^{TM} system, a unique, scientific approach to customer satisfaction, integrates Quality Strategies' considerable business expertise with leading edge methodology to maximize the ability to profitably manage customer relationships. The Action ^{TM} system goes beyond traditional Customer Satisfaction tracking studies. By linking your organization's internal work systems and processes to Customer Satisfaction intelligence, substantive product and service improvements are defined. Employees are given clear objectives to act against to satisfy and retain customers. Once enabled with practical quality tools and customer-driven training, employees can create and implement effective solutions that impact the bottom line.

Client listing for Customer Satisfaction studies

Client list is proprietary, categories include utilities, consumer, industrial, and business-to-business.

Additional capabilities (training, presentations)

The additional M/A/R/C capabilities are available to integrate into a customer satisfaction measurement program or can be used independently.

179

CUSTOMER SATISFACTION RESEARCH RESOURCE

Expertise US [X]   International [X]

Name

WILLARD & SHULLMAN

Address   325 GREENWICH AVENUE, GREENWICH, CT 06830

Phone   203-629-2233   Fax   203-629-2275

Key contact   Robert R. Shullman   Title   President

Skill summary (techniques, models, process names)

Place major emphasis on role of employees in Customer Satisfaction. Integrate employees' role and satisfaction with ability to meet/exceed customers' expectations. Customer Satisfaction research results identify *where* the issues are – employee satisfaction results uncover *how* to address issues.

Particular expertise in business-to-business Customer Satisfaction measurement.

Client listing for Customer Satisfaction studies

Aetna, Brooklyn Union Gas, Citibank, Chemical Bank, Compaq, Dow Chemical, IBM (Publications), Metpath, Nasdaq, New York Stock Exchange, Pitney Bowes, Time Inc., Warner-Lambert.

Additional capabilities (training, presentations)

Can conduct orientations, presentations, and training on Customer Satisfaction, from overview to detailed half-day, tactical sessions.

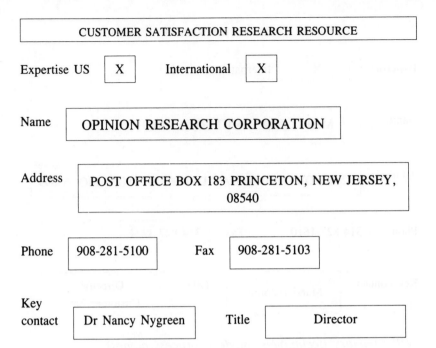

CUSTOMER SATISFACTION RESEARCH RESOURCE

Expertise US [X]   International [X]

Name   OPINION RESEARCH CORPORATION

Address   POST OFFICE BOX 183 PRINCETON, NEW JERSEY, 08540

Phone   908-281-5100   Fax   908-281-5103

Key contact   Dr Nancy Nygreen   Title   Director

Skill summary (techniques, models, process names)

Extensive experience in Customer Satisfaction measurement and organizational diagnostics. Emphasize integration of internal and external. Also customer panels, business process redesign, and implementation rollout. Extensive health care/pharmaceutical, technology, and financial services experience.

Client listing for Customer Satisfaction studies

Johnson & Johnson, AT&T, General Electric, Marriott, New York Stock Exchange, Lukens Steel, CSX Transportation, PNC Bank, Moody's Investor Service, US Postal Service.

Additional capabilities (training, presentations)

ORC has wholly owned European telephone interviewing capabilities; normative databases; advanced modeling capabilities; train-the-trainer feedback programs.

```
┌─────────────────────────────────────────────────────────┐
│         CUSTOMER SATISFACTION RESEARCH RESOURCE           │
└─────────────────────────────────────────────────────────┘
```

Expertise US   [X]   International   [X]

Name   ┌─────────────────────────────────────┐
       │   MARITZ MARKET RESEARCH   │
       └─────────────────────────────────────┘

Address   ┌──┐
          │ 1297 NORTH HIGHWAY DRIVE, FENTON, MO. 63099 │
          └──┘

Phone   ┌──────────────────┐   Fax   ┌──────────────────┐
        │ 314-827-1610   │   │ 314-827-3224   │
        └──────────────────┘   └──────────────────┘

Key contact   ┌─────────────────────┐   Title   ┌─────────────────────┐
              │   Marsha Young   │   │   Director   │
              │   │   │   Customer Sat.   │
              └─────────────────────┘   └─────────────────────┘

Skill summary (techniques, models, process names)

Business-to-business speciality; custom-designed Customer Satisfaction measurement systems; national data collection network;
actionable analysis including performance improvement planners
and vulnerability analysis, multi-industry expertise. More Than
Measurement^{TM} approach to maximize the voice of the customer.

Client listing for Customer Satisfaction studies

Federal Express, General Motors, Caterpillar, Fort Sanders Health
System, IBM, Southwestern Bell, Medtronic, Southern California
Gas Co., Waste Management, AT&T, Saturn Division of GM.

Additional capabilities (training, presentations)

Maritz Inc., a $1.4 billion performance improvement organization,
offers More Than Measurement^{TM} continuous quality improvement
strategies including deployment of information, customer-directed training, employee involvement and consultation of recognition and reward
systems relating to compensation. Maritz provides systematic approaches
to managing the customer relationship to enhance loyalty and retention
including database marketing and loyalty marketing programs.

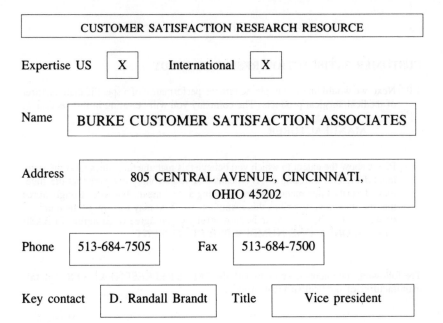

CUSTOMER SATISFACTION RESEARCH RESOURCE

Expertise US [X] International [X]

Name | BURKE CUSTOMER SATISFACTION ASSOCIATES

Address | 805 CENTRAL AVENUE, CINCINNATI, OHIO 45202

Phone | 513-684-7505   Fax | 513-684-7500

Key contact | D. Randall Brandt   Title | Vice president

Skill summary (techniques, models, process names)

Extensive experience with Customer Satisfaction measurement, part of Burke Market Research Organization with worldwide capability. View Customer Satisfaction from internal (employees) as well as external perspective, with special emphasis on 'chain of customers' such as distributors, etc. Extensive professional–health care experience.

Client listing for Customer Satisfaction studies

Abbott Labs, AT&T Microelectronics Div., Bellsouth, Eli Lilly, First Nationwide Bank, Hewlett-Packard, Kraft Foods, Microsoft, Official Airline Guides, Ralph's Grocery, Sea-land Corp., Steelcase Corp., Syntex Lab., Westin Hotels, Weyerhaeuser.

Additional capabilities (training, presentations)

D. Randall Brandt is a frequent speaker at Customer Satisfaction seminars, presentations, and conferences. Can also provide training in Customer Satisfaction measurement and develop reward/ recognition systems related to Customer Satisfaction.

CUSTOMER SATISFACTION RESEARCH STUDY

4.0 Next, we would like you to please rate the performance of a specific manufacturer of medical/surgical products. The company you will be rating in this section is:

MANUFACTURER _____

Please show the extent to which you believe this manufacturer does what the statement describes. Circling a ''1'' means that you strongly disagree that the statement describes this manufacturer. Circling a ''7'' means that you strongly agree that the statement describes this manufacturer. You may use any of the numbers between ''1'' and ''7'' to show how strongly you agree or disagree. PLEASE CIRCLE ONLY ONE NUMBER PER STATEMENT.

The following statements have to do with the SALES REPRESENTATIVES from this manufacturer of medical/surgical products.

	Strongly Agree	Strongly Disagree	
This company's sales representative follows my hospital protocol for when and who they can call on	7......6......5......4......3......2......1		(B07)
This company's sales representative makes personal contact with me twice a month	7......6......5......4......3......2......1		(B08)
This company's sales representative informs me of special pricing or discounts with enough time for me to take advantage of them	7......6......5......4......3......2......1		(B09)
This company's sales representative knows his/her product well	7......6......5......4......3......2......1		(B10)
This company's sales representative understands materials management personnel needs and views them as part of the team	7......6......5......4......3......2......1		(B11)
This company's sales representative develops programs to reduce total costs for my department and hospital, other than unit price reduction	7......6......5......4......3......2......1		(B12)

184

QUESTION 4.0 CONTINUED

	Strongly Agree						Strongly Disagree	
This company's sales representative responds quickly (within 24 hours) when called and follows up in a timely manner	7	6	5	4	3	2	1	(B13)
This company's sales representative offers suggestions on how to improve my current inventory management	7	6	5	4	3	2	1	(B14)

The following statements have to do with PLACING THE ORDER with this manufacturer of medical/surgical products.

	Strongly Agree						Strongly Disagree	
This company has Automated Order Entry (EDI) capability	7	6	5	4	3	2	1	(B18)
This company has fax as an option for sending in orders	7	6	5	4	3	2	1	(B19)
This company has an order process which makes placing an order for me easy	7	6	5	4	3	2	1	(B22)
This company provides confirmation of orders that meets my needs	7	6	5	4	3	2	1	(B23)
This company has customer service representatives who are able to answer my technical and clinical product questions	7	6	5	4	3	2	1	(B20)
This company has customer service representatives who are able to answer my general business questions (such as the price of the product)	7	6	5	4	3	2	1	(B21)
This company has customer service representatives who know my account history and personnel	7	6	5	4	3	2	1	(B24)
This company has customer service representatives who can answer my questions without having to transfer my call	7	6	5	4	3	2	1	(B25)

185

QUESTION 4.0 CONTINUED

	Strongly Agree						Strongly Disagree	

This company has customer
service representatives who are
available when I call ....7......6......5......4......3......2......1   (B26)

This company has customer
service representatives who warn
my department of a credit hold
before placing an order..................7......6......5......4......3......2......1   (xxx)

The following statements have to do with the HANDLING OF BACK ORDERS at this manufacturer of medical/surgical products.

This company's customer service
representative tells the person
placing the order immediately if
product is on back order ....7......6......5......4......3......2......1   (B28)

This company's customer service
representative tells me when
back ordered product will be
available ....7......6......5......4......3......2......1   (B29)

This company's customer service
representative is willing to find
substitutes of that company's
products when back ordered
product is not available.................7......6......5......4......3......2......1   (B30)

This company's field sales
representative is willing to find
substitutes of that company's
products when back ordered
product is not available ....7......6......5......4......3......2......1   (xxx)

This company provides a fill rate
that meets my standards ....7......6......5......4......3......2......1   (B32)

The following statements have to do with the ARRIVAL OF THE PRODUCT from this manufacturer of medical/surgical products

This company delivers the
product when promised ....7......6......5......4......3......2......1   (B33)

QUESTION 4.0 CONTINUED

	Strongly Agree						Strongly Disagree	
This company delivers the product during the hours my shipping dock is open to receive orders	7	6	5	4	3	2	1	(B34)
This company's shipment is palletized	7	6	5	4	3	2	1	(B35)
This company monitors the performance of the carriers they use and share this information with me	7	6	5	4	3	2	1	(B36)
This company ships product so that it arrives within 3 days of order placement	7	6	5	4	3	2	1	(B37)
This company ships my standing order so that it arrives on my pre-scheduled days	7	6	5	4	3	2	1	(B38)

187

Index

189

191